# Memoir of a Flight Attendant

Margo Deal Anderson

Memoir of a Flight Attendant

Copyright © 2014  Margo D. Anderson

All rights reserved.    LCCN 2016901551

ISBN:
ISBN-13: 978-1499235609

# Memoir of a Flight Attendant

## DEDICATION

This memoir is dedicated to the flight attendants who never returned home on 9/11; they will forever be in my memory and in my heart.

Memoir of a Flight Attendant

Memoir of a Flight Attendant

Memoir of a Flight Attendant

## ACKNOWLEDGMENTS

The events described in this memoir are based on true events; names have been changed to protect the innocent (and the guilty). I am proud to have had the opportunity to be a flight attendant, and I hope if any of my former colleagues read this memoir they will recognize themselves, remember, and smile. Especially Vicky and Ruth.

**To my husband, Lee,** *who I love with all my heart, and who proposed to me on 9/11 as soon as I stepped off my flight.*

**To my beautiful daughter, Hilary,** *who inspired me to finish this memoir.*

**To my granddaughters, Lily and Natalie,** *who I hope will always see their Mimi as adventurous and unafraid.*

**To my brother, Roy,** *who has always been my best friend.*

**To my Mom,** *without whom I would never have survived the hurdles of my life.*

Memoir of a Flight Attendant

Memoir of a Flight Attendant

# 1  A Ferret

The little boy was charming in an Opey from the Andy Griffith show kind of way; he carried his pet cage on and smiled at me as he pulled his Transformers rolling bag behind him.  His mom ignored him as she adjusted her sunglasses and looked at me disdainfully; his dad looked tired and called to him to "wait up, buddy…," which he promptly ignored and said, "I can find my seat, it's 6A."

"Smart kid," I thought.  I watched as he got into his seat, pushed his cat carrier under the seat, petted the animal, and promptly settled in with his video game, oblivious to the fact that he was in the aisle seat and that his mom would have to climb over him. Dad looked relieved as he took the seat across from them in the small, fast, commuter jet.

I was particularly happy today because I was flying the forty-seat version of the Canadair jet; ten less passengers to deal with than the regular fifty- seat design, more room in the closet for my stuff, and it was the last day of a grueling four day trip. No stress today, we had just come in from Asheville, North Carolina, a city I fondly referred to as the gray people; they really looked gray most of the time. Maybe it was the cloudy, gray days which seemed to prevail most of the time when I landed there, or maybe it was just the age of the average passenger: gray beards, women who chose not to color their hair, a kind of leftover from the sixties, liberal, earth-sandal group of people. At any rate, they were generally a no problem group as far as I was concerned. They wanted hot tea with no cream or sugar, black coffees, or fruit juice with no ice. They never wanted the snacks because they only liked whole-grain stuff, not the white-flour pretzels. The most exciting thing that ever happened on that flight was if Andie McDowell (who lived in Asheville ) was on the flight; she would sign autographs for passengers and smile disarmingly, unusual for someone with her level of stardom...God, I loved *Four Weddings and a Funeral*...not to mention the fact that this woman had actually touched Hugh Grant.

    I had just finished the unbelievably long memorized safety announcement which ASA required we make on each flight and had begun checking seat belts, overhead bins, reminding people to make sure their luggage was completely under the seats, and collecting Starbucks cups , salad plates and Cinnabon boxes, when the Captain rang the familiar "ding-dong," the two tones which indicate "cockpit calling."

    I answered the phone to see what the crew needed, and the First Officer let me know that we had a maintenance issue--- it would be a few minutes before we would be leaving; would I mind bringing him some coffee, black? No problem, I always wanted my flight crew to be relaxed *and* awake.

"*God, I hope we make up the time on the way home; I really want to make my flight home to Panama City,*" I thought to myself.

The maintenance guys came aboard and went into the cockpit; nothing makes the passengers feel secure like guys with toolboxes entering the cockpit. The Captain had not yet made his announcement to let passengers know of the delay, and so immediately the questions started. I told them it was all about "some little light" that didn't work properly in the cockpit and as soon as they had it fixed we would be on our way. With my long blonde hair I could get away with ridiculous statements like this.

Immediately all of the laptops, headsets, and other portable electronic devices came back out again, and the passengers began calling on their cell phones to let friends and families know they would once again be delayed in Atlanta.

"ASA-always sitting in Atlanta," and, "I'm stuck in Atlanta again with ASA," I heard passengers commenting.

As I made the fresh coffee for the crew I looked around at the passengers. In Row C was an obese couple who seemed to melt into each other, and they were finishing a value-sized meal from the airport Wendy's (OMG); Row D held four corporate guys dressed in matching computer logo polo shirts with blue tooth devices on their heads and computers in their laps....I remembered the geeks in the movie, *Sixteen Candles*. "*All they need is headgear and retainers,*" I thought.

In Row One was the usual asshole who always puts his carry-on on top of my stuff (which I always put in the first overhead bin), takes off his shoes and props his nasty, smelly feet on the bulkhead. The forty-seat regional jet has a great front row with tons of legroom. Whoever sits in this row always assumes this is the commuter jet equivalent of First Class and begins to whine as soon as I tell him or her that, " Yes, the cocktails are $5.00,"and "No, I do not have change for $100.00." Back to the asshole, he also usually brings a copy of either the *Wall Street Journal* or *New York Times,* which he scatters all over the floor and leaves for me to pick up. This is second only in aggravation to the parents who leave Pampers under the seat filled with baby's latest surprise.

Fifteen minutes later the maintenance guy emerges from the cockpit, sweaty, smelly and disgusted, hands the crew the pink copy of the paperwork, and as he leaves asks me if I have a bottle of water I can spare...of course I do...I just want to close the door and leave Atlanta for the second time today. I close the door, lock the cockpit door, secure the coffee pot, and start the cabin walkthrough...again. As I remind passengers to secure their items underneath the seat in front of them, turn off portable electronic devices, and take their drink cups, I notice the cute little Opey fellow in Row Six is leaning forward talking to his cat, whose little pink nose is poking through the grid on the under seat cat-carrier.

"Everything ok with kitty?" I ask.

"He's chillin," Opey answered.

I sit down, fasten my seatbelt, pick up the phone and tell the crew we're secure for takeoff. Twenty minutes later, we are still crawling along the runway, waiting in line for our turn to leave Hartsfield Airport. Finally, the Captain rings the bells to let me know that we are about to take off, and I smile in anticipation of two more legs—a round trip to Arkansas, this day will be over, and I will be running through Concourse C to catch the late flight into Panama City to see my boyfriend Lee. *"Home for four days. I can't wait."*

I feel the familiar acceleration, and I say my usual mantra: "Lord keep us safe through this takeoff and flight." All flight crew members have their little things they do during takeoff; we may look calm, but after several thousand takeoffs you know that eventually the law of averages may be working against you. I try not to think about it, but I have my preferences like everyone else. I don't like flying with athletic teams, newlyweds or famous bands. They are all poison to flying Karma. I know it; they know it; we all know it.

We make it through the takeoff and as I feel the Canadair commuter jet continue to climb toward our first goal of 10,000 feet I feel myself begin the slow exhale of "thank-you" to God, angels and other elements of Karma which have allowed me to safely take off for flight 4,350 whatever it is. Now that all is well, I sneak out my most recent issue of Cosmo and begin to read behind my flight attendant manual for the next ten minutes or so before the passengers will begin to look expectantly at me for their cups of Diet Coke, fruit juice, Bloody Mary mix, and whatever the Hell else they can think of to order to make life difficult. Diet Coke is the most aggravating drink to pour --- the altitude makes it foam even more than usual. The ladies who want hot tea irritate me because I have to walk all the way back to the galley to get the hot water which breaks my serving rhythm. They usually wait until we are in some kind of turbulence to order hot tea just to see if I can't get a third degree burn trying to serve them.

This flight was going to be different, however. Just as I stood up and pulled the cart out to begin my journey down the aisle of passengers, I heard a blood-curdling scream.

"What the fuck!" yelled the asshole in the front row as he turned to look over his shoulder. Just at that moment a streak of fur went over his seat , across the aisle, and seemingly flew over the next several seats on the right side of the airplane.

I immediately pulled the cart back, locked it in place, and cautiously peered out at the passengers. Two more screams, "What the hell was that?", and a "Holy shit!" greeted my ears as I started up the aisle. Opey was out of his seat, his mom was yelling at him, and his dad was headed toward the back of the aircraft near the head. Four or five passengers were getting up headed toward me, when I saw it. A long, furry rodent streaked across the airplane, went under a seat and then reappeared on the back of the seat of the obese couple. The Wendy's bag went straight in the air and the cabin was raining French fries.

"It's a rat!" screamed the obese woman.

"It is not," said Opey , "it's my ferret."

"Oh my God," I shouted.

" I thought you had a cat!"

"No, it's Fernando---my ferret. They're gonna hurt him."

I ran back to my telephone and picked it up.

"Ladies and gentlemen, please remain in your seats; this little boy has lost his ferret. I will help him catch it. Please remain calm."

No one was calm. As I looked about the cabin newspapers were swatting at the animal, feet went straight up, drinks flew to the ceiling, and the screams grew louder.

"Ding-dong, Ding-dong," the Captain was calling.

" Great," I thought.

"What?" I said as I picked up the phone. "I have a ferret loose in the cabin, and people are losing it."

"Say again," said the Captain.

"A fucking ferret is running all over the cabin and I am trying to catch it."

Silence.

"Call me back when you have it."

I hung up.

"Everybody sit down," I shouted. "It's just a ferret."

"It's a goddamn rat," shouted the asshole in the first row. He picked up his briefcase and stood up. " I'll kill the little sonofabitch," he said.

"Sit down," I shouted. He sat down.

"Get the cage," I yelled at Opey. He grabbed the carrier from beneath the seat and followed me toward the middle of the cabin where the ferret was now standing on its hind legs, with its nose twitching inquisitively, climbing onto the armrest of a seat, staring at the horrified woman who had dropped her cup of hot tea.

I snatched a blanket from the lap of an older man who for some reason was still napping, and I began moving stealthily toward the ferret.

"Don't hurt him," said Opey.

"Shut up ," I whispered, "I'm not going to hurt him...open the door of the cage."

Just as I lunged toward the ferret with the blanket it disappeared under the seat, and the woman jumped up and threw a glass of juice and pretzels all over me and the gentleman next to her.

"Nice," said a teenaged kid sitting across the aisle. "Where'd he go?"

Opey was under the seat calling the ferret, and people four rows back were standing on their seats.

The familiar "ding dong" of the Captain calling sounded.

I ran to the front of the plane looking back as I ran.

"Get him," I snarled at Opey.

"What?" I snapped, as I picked up the phone.

"It sounds like all hell has broken loose back there; how hard is it to get a goddamn ferret in a cage?"

"I'll call you back," I said.

"*Screw you*," I thought as I slammed the phone down on the hook.

Opey had the ferret in his hands petting it when I turned around, and the passengers were applauding.

I held the cage while he put the squirming little animal, nose-first through the door, and I slammed it shut.

I helped Opey back to his seat, shoved the cage underneath the seat in front of him, and whispered to him, "If the little bastard gets out again, I will step on him, understand?"

Opey nodded at me solemnly with his eyes wide open.

 "What did you just say?" asked his mom.

I said, "I certainly hope the ferret is ok; please make sure he stays in the cage."

"You are aware that airline regulations do not allow rodents in the cabin?" I continued to the mom. " Only cats and dogs…well-behaved cats and dogs."

I turned and walked to the front before the glaring mom could answer; his father continued reading his book without comment…just a dirty look at me.

I picked up the phone and rang the cockpit.

"Ferret is secure."

"Could I have some orange and cranberry juice with no ice, mixed?" asked the First Officer.

"No," I answered, and hung up.

I sat down in my jump seat and began reading my Cosmo magazine, which is strictly against company policy…with everyone watching.

Memoir of a Flight Attendant

## 2 MONTERREY

"Today we will be training for the upcoming flight destination of Monterrey, Mexico," said our flight attendant instructor.

I was excited about the fact that our small regional aircrafts would now begin making international flights to Monterrey, Toronto, and Montreal. Anything would be a relief from the short flights from Atlanta to such exciting destinations as Montgomery, Augusta, Wilmington, Asheville, Dothan, and Panama City.

As the instructor began outlining all of the pertinent information and warnings about international travel, I began to hear my inner sarcastic voice speaking: *" Really, if we leave our bag unattended in the airport they will take it to the incinerator and blow it up? Please. "*

We were told we would each be allowed a case of water to remove from the airplane for drinking and washing our hair if we wished.

"*Why?*" I mused. "*We are staying in the Doubletree resort in Monterrey….they serve frozen margaritas and warm chocolate chip cookies. Don't eat the food or drink the beverages?* "

" *Ok,*" I thought. " *You stay in your room and drink bottled water and pretzel snacks from the airplane…I'm going downstairs and have a frozen margarita and some burritos.*"

After daydreaming through the rest of the Monterrey training, I decided I was definitely going to bid for the trip because no one else seemed interested after the scare tactics of the training class. A three- hour leg on the forty- seat jet with a twenty-four hour layover at the Doubletree sounded great. "*Sign me up. Cool….I'm going,*" I decided.

I bid the trip, and I got the trip. " Zip uh dee doo dah….Margaritaville here I come!" I exclaimed happily when I saw the bid announcements for the month.

As the passengers boarded the flight for my second trip to Monterrey, the first thing I noticed was that all of them were bitching.

"You've got to be kidding me…"   "What is this, a Barbie jet…", "No First Class, I've got to PAY for drinks?"

The passengers were pissed because they were accustomed to flying the larger Delta jets to Monterrey, and now they had been "downgraded" to Atlantic Southeast Airlines, Delta Connection, for the three- hour flight.

"Where the hell do I put my carry-on?" asked the tall, nice-looking gentlemen as he came through the door, bumping his head and swearing at the same time.

"In the overhead compartment," I answered smartly.

"What if it doesn't fit?" he demanded glaring at me.

"Then I can check it for you," I retorted, glaring back.

Without answering, he plopped down into the first row of the forty- seat regional jet; there is a lot of legroom in this bulkhead seat, and he left his bag sitting in the aisle for me to stow.

I picked it up and shoved it into the overhead compartment.

"See," I said to him. "It fits!"

"Bring me a Scotch and soda," he ordered.

"Sorry, sir. We can't serve any alcohol until we are above ten thousand feet. It is the international law regarding service of liquor," I explained.

"You're kidding," he said.

"Sorry, company policy."

"This is crap," he said. "Wait until I tell Leo how I feel about this."

I ignored him; Leo was the CEO of Delta at that time, and I really doubted that this jerk, like about one hundred others who make the same threat on an almost daily basis, actually knew him.

I walked to the aft of the aircraft, smiling sweetly at the irritated passengers, reminding them to fasten seat belts, raise seat backs, and, " yes, bags have to be stowed completely underneath the seat in front of you."

"But I like to rest my feet on my bag," whined the lady in 5A. "What can it hurt?" "Besides, it won't fit under the seat or in the overhead."

"It's for safety reasons," I reminded her, in my most businesslike voice. "

*"So that you will be able to get out of your damn seat and off this airplane if we have to leave in a hurry,"* I thought, *"and if you hadn't been such a know-it-all, unpleasant bitch, you would have let me put a pink gate tag on it and you could have retrieved it at the gate….now you will have to go to baggage claim to get it since I no longer have pink tags and I will have to hand it out the door to the ramp guys."*

"I am so sorry, ma'am; I will take it for you, and you can pick it up at baggage claim in Monterrey," I said as I pulled the bag into the aisle and began wrestling it to the front of the aircraft.

"Baggage claim," she cried, "this is ridiculous!" "You can bet that I will be writing a letter to Delta about this."

"There's a comment card inside the magazine in the seat pocket," I said as I smiled sweetly at her.

I returned to the galley area to retrieve my seatbelt and oxygen demo equipment and got ready to begin my long memorized safety announcement….Leo's best buddy in the front row had now removed his Gucci loafers and had his feet propped on the bulkhead revealing that he wore long dress socks held up by what appeared to be garters.

*"Oh my God,"* I giggled to myself.

I called the cockpit and said, " Thirty-six passengers guys, no kids."

"Got it," said Jonathan, the First Officer.

"You guys need something to drink, or one of the little salami and cheese food packs?" I asked.

"Michael wants one," said Jonathan, "but I don't. Those are fart-makers."

"Gross," I answered.

The Monterrey flight had initiated a little snack pack which included several gross items, including a miniature summer sausage, cheese and beef jerky. They were popular with the pilots, but NOT the flight attendants. The passengers were even less impressed.

The guys handed me the paperwork; I handed it out to the ramp guys and began closing the forward door. I then closed and locked the cockpit, secured the galley, and started my final walkthrough, checking seatbelts, overhead compartment doors, and making sure trays and seatbacks were up.

I made *no* eye-contact with the passengers as I completed this task, ensuring that they couldn't ask me to do anything before takeoff. After completing the safety announcement and demonstration, I sat down, clicked the seat belt on the jump seat, and simultaneously picked up the phone to let the guys know I was ready.

"Cabin secure for takeoff," I said.

"Roger that," answered Jonathan.

I watched out the galley window as they made the turn to line up for takeoff at Hartsfield.

"Ladies and gentlemen, we are currently sixth in line for takeoff," said Jonathan. "Sit back, relax, and enjoy your flight."

Asshole glared at me, and I quickly looked away.

*"Sissy with sock-holder uppers,"* I thought.

I thought with anticipation of twenty-four hours to relax in Monterrey.  The first time I went there I couldn't wait to get out and walk around and shop at the little stands and shops near the Doubletree, but this time I was looking forward to staying in the room and sleeping.  I had been flying for three days already, and after this overnight in Monterrey I would be home for almost eight days.  I was so excited.

I felt the plane accelerate, and I quickly said my ritual takeoff prayer, and then looked at the passengers, planning an escape plan if anything went wrong.  I often wondered how many flight attendants did this, but I always had in mind who I would call upon to help me, who would need assistance, and who could take care of themselves in the event of an emergency.  In just a few seconds I felt the sleek Canadair regional jet gain altitude with its familiar speed and power, and then just as quickly I heard the "dings" from the cockpit signaling that we had reached 10,000 feet.

Most of the passengers were business people, familiar with the routine, and almost before the bells finished chiming, they had their laptops out, working away.

Not the asshole.

"How about a cocktail now, or is there a rule about *how many* feet we have to be in the air before you can serve?"

I smiled my most pleasant smile and answered, "It will take me just a moment to get the cart out, and I will be happy to take care of you."

He didn't answer, and I began putting ice in the ice bucket, wrestling the beverage cart out of its cubby in the galley, and positioning it to move up the aisle. I normally pushed the cart to the aft, or back of the aircraft to begin serving, working my way back to the galley, but I decided that today I would start from the front , in hopes of avoiding further confrontation with Mr. Delta, the Platinum Medallion passenger of the century.

"What can I get for you?" I asked him.

"Vodka tonic with a lime," he answered.

"I'm sorry; I can take care of the vodka and tonic but we don't have fruits aboard."  *"Except you," I thought.*

"Are you kidding, me?" he answered.

"That will be four dollars, sir," I replied, ignoring his question.

"Four dollars?  Do you realize that I am a platinum medallion member?"

"Congratulations.  It's still four dollars."

"I don't like your attitude," he said.  "Besides, I only have a hundred dollar bill on me and you girls never have change so I guess I'll get it free anyway."

*"This is too choice for words,"* I thought.  Most of the experienced passengers told us this, because they knew if we didn't have change we would have to comp the drink.  I had gotten so tired of this because I had to pay for the drinks out of my own money when I didn't have change in order to make the liquor kit balance

for catering in Atlanta. I had gone to the bank and gotten one hundred *one* dollar bills, and I had been carrying them around for about two weeks, just waiting for an opportunity like this. Thankfully most airlines accept only credit cards now.

He waved the one hundred dollar bill in the air, and I neatly clipped it from his fingers in one quick swoop.

" No problem, I have change, " I said, and I began quickly counting out ninety-six ones into his lap.  He was speechless for a few seconds, and then said weakly, "Don't you have any larger bills?"

"Nope," I said, and began pushing the cart to the next row.

"What would you like to drink, sir?" I asked the thin, tired- looking man behind him, laughing silently to myself over my small victory with the asshole.

"I'll have an orange juice with no ice," he answered, looking out the window.

I love the juice with no ice orders-- quick, no foaming up to wait on like diet sodas, no mess.

I continued quickly through the cabin, watching with fascination as each passenger turned the snack packages around, peering through the wrappers in disgust at the contents.

When I reached the last seat, I put the brake down on the cart and quickly scooted into the restroom, knowing I was breaking the rules leaving the cart unattended, but I was about to pee my pants.

I sat down quickly on the seat of the toilet.

 "Ugh," I said in horror.  The last passenger had wet the seat.

"Oh, my God," I said out loud. "How nasty can people be?"

I grabbed a paper towel and wiped off my wet butt, wondering how many germs had just touched my bare bottom.

"I hate these damn toilets," I said out loud to myself.

I came out of the head, a determined smile plastered to my face and began making my way back to the galley, picking up empty cups and asking passengers if they would like anything else. The new snack packs were 80% wrappings and one trash bag would barely hold all of the trash, plus newspapers, and in the case of several smarter passengers who had brought their own drinks and food, I had to make room for Cinnabon boxes, Burger King bags, and large Starbucks cups.

"You know that I am going to report how you have treated me," said asshole as I returned to the galley in view of his front row seat. "And you can bet that I will never fly on this miniature Barbie jet again."

*I had had enough.* I turned to him and looked directly into his eyes and said in a loud voice, "I don't care who you report me to; if you are so rich and entitled that you know Leo personally, why don't you buy your own damn thirty million dollar jet like this one, and then you and he can fly privately to Monterrey, stock your own lime slices and have all the free drinks you want."

I then turned my back to him and began finishing my cleanup work in the galley. He was either too stunned or too angry to say anything back, and I thought, *"Holy shit, this will probably be my last flight because I will be fired when I get back to Atlanta."*

Since I knew this was my last flight, I sat down in the jump seat and began to read a magazine. As I thought more about this, I put the magazine down, went to the closet, grabbed my small lunch box and set it on the counter in the galley out of sight of the passengers. I took out my chicken soup mix, sprinkled it with salt and pepper in a drink cup, and mixed it with hot water from the spout near the coffee maker. Soon the cabin filled with the smell of chicken; immediately I heard the familiar "ding" sound of a passenger call button. I peered around the bulkhead and saw the gentleman in Row Nine looking at me expectantly.

I walked back to him. "Something wrong?" I asked.

"No," he said. "I just smelled the chicken and wondered when lunch would be served."

"I'm sorry," I apologized. "This is not a meal flight; we only serve the enhanced snacks."

"Well, what's that I smell cooking if this is not a meal flight? Special dinner for the pilots?" he sneered.

"No; actually, it's my lunch. Instant chicken soup from an envelope mixed with hot water."

"Oh." "It sure smells good," he said.

I pressed the flight attendant call button light off, turned quickly and returned to the galley and my soup.

I opened a Diet Coke, poured it over ice, and sat on my flight bag with my back against the wall near the emergency door where no one could see me.

"*I hate them,*" I thought. "*I don't care if they do fire me.*" I propped my feet against the galley cart and looked at the navy blue pumps on my feet. *"It will be nice to wear something besides navy blue."*

I looked out the port hole on the emergency door and saw only the tops of beautiful cotton-ball like clouds with the afternoon sun shining brightly against them. This was the part of flying that I really loved; after the passengers were served, on a long flight like this one I could sit and enjoy the absolute beauty of being in the air.

Soon the mountain-tops of Monterrey, "ciudad de montanas," the city of mountains, began to loom in the distance, and I admired as always the skill of the crew as they began the initial approach which would bring the regional jet through the clouds and mountains of Mexico to the beautiful city of Monterrey.

Memoir of a Flight Attendant

## 3  Flying the Plane

"Well, it looks like we have an empty plane going back to Atlanta tonight," said Sheila the other flight attendant on the ATR.

This was not an unusual occurrence right after September 11, 2001.  Although our company resumed its regular flights on September 16th, the flying public was not so sure.  Flight after flight would be filled with only the crew—flight attendants and pilots.

"I'm certainly not disappointed," I answered.  "I think I will relax, eat my supper, and be ready to change clothes and run for the flight to Panama City."

 For once,  my flight home was not filled with tourists, spring breakers, or athletes headed to an Iron Man competition—the employee computer web page showed forty-two empty seats on the sixty-six seat plane going to my hometown.  Panama City is such a popular tourist destination in northwest Florida that getting home or to work was sometimes extremely stressful.

"I'm not even hungry," answered Sheila.  "I'm going to get in the backseat, stretch out, and take a nap."

"Leave the cockpit door open if you want," said Patrick, the Captain. "It's not like anyone's going to storm the cockpit at 18,000 feet," he laughed.

I loved flying with Patrick; he always treated flight attendants as valuable crew members instead of like "waitresses in the sky" like some of the pilots did. He wasn't full of himself, didn't talk about his $100,000 degree from Embry-Riddle, and when he wanted coffee he fixed it himself as he came aboard.

"Ok, " I answered, "let me know if you need anything."

"Could I get some coffee up here when you have time," asked the First Officer.

"Sure, " I answered. "What was your name again?"

"Hans," he said, in a thick German, maybe Scandinavian accent.

"Cream, sugar, black?" I asked.

"Black," he said.

I made the coffee, took it up to the cockpit, and returned to the back of the aircraft. The ATR is an ugly, sixty-six seat aircraft, manufactured in France, and I absolutely hate flying on it. It looks like a gooney bird, heavy in the bottom, wings set above the body of the plane—I always wondered how it would possibly get off the ground during takeoffs. ASA bought these planes to use in the southeast because they have a real ice problem with the wings in more northern climates. The galley is in the back of the plane, and if you are the "B" flight attendant, your jump seat is in the front with a hallway connecting to the pilots in the cockpit. If the light is not on in the hallway, you walk through black dark to get to them, with luggage on both sides of you. Also, something is ALWAYS broken on the plane, and maintenance is a nightmare. The only cool thing about the aircraft is that the rear baggage compartment is accessible from the galley, and so flight attendant bags can be tossed back there if the overhead compartments are filled on the plane. Before security became so tight after 9/11, we also used the compartment for taking naps on top of the luggage or hiding crew members trying to commute home when all the seats were full and we could get away with it.

Tonight's flight was extremely short, Augusta to Atlanta, and we had enjoyed a leisurely eighteen- hour layover in Augusta at one of our favorite hotels, The Partridge Inn, a great bed and breakfast boutique hotel. As we reached 10,000 feet, Patrick called me on the phone, and I answered, "What?"

"Can you come up here for a minute?" he asked.

"Why?"

I had just settled in with a cup of hot soup from my thermos and my Marie Claire magazine.

"Because I need to show you something."

"Ok." I laid the magazine down, set my soup on the adjacent tray table, and with some degree of irritation walked to the front.

"*What in the hell do they want now*? " I mused to myself.

I noticed that Sheila was already stretched out across three seats with two pillows and blankets taking a nap. I walked through the hallway between the cabin and the cockpit which was also loaded with baggage on either side, and I could see the familiar glow of the cockpit control panel lit against the dark night visible through the windshield.

"Would you like to fly the plane?" Patrick asked casually.

"What? Are you kidding?" I exclaimed.

"Are you crazy?" said Hans, as he sat up a little straighter in his seat.

"I said, would you like to fly the plane? " Patrick repeated with a little gleam in his eye.

"Absolutely," I answered.

"Oh, goddamn," cursed Hans, "we all gonna get fired; no fucking way!" He sounded like Arnold in the Terminator I thought to myself.

"We are not going to be fired," Patrick answered calmly.

"Get out of your seat, and get in the jump seat," he directed Hans.

"Margo, you know how to use the pilot oxygen mask, don't you?"

"Yep, " I answered. We had to learn to use the cockpit oxygen after 9/11 because the new safety regulations required one of us to go in the cockpit if a pilot came out to go to the head.

Hans got out of his seat, cursing under his breath, and I excitedly climbed into his seat and began fastening my seat belt as he angrily pulled down the jump seat which is used for pilots who are flying off duty, FAA personnel, or flight checks.

"You know this is all recorded; they are going to listen to it, we all gonna get fired," Hans repeated in his thick accent.

"They will only listen to it if we crash, and then it won't matter," Patrick answered grinning.

"Great," said Hans.

"Ok, I'm taking it off auto so that you can really feel the plane," said Patrick.

"No!" said Hans.

"Cool," I said as I put my hand tightly on the yoke. The craft felt awkward and heavy, and I was tense. "This is not easy," I thought to myself, wondering how actress Karen Black flew the 747 in the vintage movie *Airport*.

"Watch the altimeter," said Patrick.

"I know, the little wings," I answered.
"Are you kidding me?" shouted Hans. "It's not the little wings!"

The aircraft began to slide; I was not doing a great job, and Patrick directed me to turn back to the right in a calm, rather amused voice.

The radio came on and ATL wanted to know something about the position; apparently we were already getting close. Augusta to Atlanta is a very short flight. Patrick spoke calmly to them while Hans, terrified, continued to make comments about losing his job, and then about losing his life.

"It's really not a bad idea to at least have some minimal flying experience in case of a real emergency," Patrick commented.

"I love this," I said excitedly.

"She loves it." said Hans. " Great, I do not believe this."

"Hold on to it," answered Patrick, "you're sliding again."

"You're sliding again," Hans echoed glumly.

ATL air traffic control began speaking with Patrick about approach, and Hans started getting out of the jump seat, preparing to take back his right seat, as the FO's seat is called.

"Sit tight," said Patrick. "Margo's gonna land the plane."

"*What?*" Hans and I exclaimed together.

Patrick had his hand over mine the whole time, and of course his feet were on the pedals, but we still bounced the plane higher than most inexperienced pilots.

"Fantastic," said Patrick.

"Thank you God, " breathed Hans.

"What a great night!" I thought as I bounded down the aisle to get my bag and head for PFN, Panama City, Florida, now known as ECP.

"No one will ever believe this," I said to Patrick.

"You are right, because it never happened," he grinned.

I never told a soul until now.

Margo Deal Anderson

Memoir of a Flight Attendant

## 4 NEBRASKA ROUND TRIP

"Damn, it's cold," said the First Officer as we climbed the steps to the regional jet.

"My feet are frozen," I answered, dragging my flight bag, lunch bag, and purse, while trying to pull my full-length navy blue wool coat tighter around me.

"What was I thinking to bid a Nebraska trip in December?" he asked.

"If you were like me, you were trying to bid a trip with Christmas off, regardless of the consequences," I replied glumly.

We finally reached the top of the stairs, began stowing our gear in the closet, and as I began making my cabin checks, he shook my hand and said, "I'm Scott."

"I'm Margo, " I answered. I felt sorry for him as I watched him go back down the stairs to do his walk-around. First Officers have to do the outside "walk-around" of the airplane as part of the checklist.

As Scott was leaving the cabin, I saw the Captain beginning the climb of the steps leading to the cabin.

"God, I hate cold weather," he said. "Hi, I'm Mike," he said.

"Margo," I answered, and smiled at him. "Would you like some coffee or hot chocolate?"

"Black hot coffee would be excellent," he said. He stowed his bag in the overhead, and then he disappeared into the cockpit.

The wind was getting colder and colder, and a light mist began falling. I finished the checklist, took Mike his coffee, and pulled the curtain back to peer outside. I saw the passengers coming down the stairs from the terminal, clutching their bags, their heads bent into the wind as they hurried toward the aircraft.

"If this keeps up they will start de-icing soon," said Scott, as he entered the cabin, cold air blowing in behind him.

"Oh, no," I said. De-icing was practically a confirmation that our flight would be delayed. Atlanta was busy enough at this time of the year without a long line of aircraft in line to be de-iced.

The first passengers entered the cabin, and I greeted them with my best flight attendant smile, hoping this would be a pleasant, non-eventful turnaround to Omaha and back. I just wanted to go home. If we were on time getting back, I could get the last flight to Panama City, and right now it had thirteen open seats. Yes!

"You've got to be kidding me!" exclaimed a huge man coming through the door, and as he bent to enter, he still hit his head on the door frame. "What the Hell is this, a toy jet?"

I cannot tell you how many times I hear this same question, and the more I hear it the more sarcastic I tend to become; although I hated ASA for a long time, I had actually begun to become quite happy with most of my colleagues, and I felt somewhat defensive about the Canadair jet, which was now my favorite ride.

"Oh, I am so sorry. NO, it is not a toy jet; it is a Canadair regional jet and it is actually quite fast."

I thought to myself, *So therefore, your big , rude ass won't be uncomfortable for long.*"

The flight was completely full, and most of the passengers were in a great mood; Santa hats, kids carrying gifts, the smell of hot chocolate and coffee, and the cold, iciness outside enhanced an overall feeling that Christmas was definitely in the air.

"You will need to push your bag completely under the seat in front of you," I reminded an old guy in a yellow sweater vest.

"It's under the seat," he replied.

"I'm sorry, but it really has to be *completely* beneath the seat," I repeated as I looked at the large duffel bag with over three-quarters of it under his feet which were comfortably propped on top of it.

"Why?" I'm not taking anyone's space, and my feet are more comfortable up," he said.

If I had a dollar for every time I had to explain this, I would *not* be flying for a living , but once again I began explaining the details of why one must stow baggage completely.

"You see, sir," I began patiently, "if we had to suddenly evacuate the aircraft for any reason, your bag would hinder your ability to get out, as well as the nice lady sitting beside you next to the window. Stowing baggage in the overhead or completely underneath the seat is an FAA regulation, and I can lose my job, or worse you could be hurt during an emergency."

"Alright, alright. I'm sorry; no one ever told me that before. I thought it was just a power thing for you stewardesses."

He began kicking his bag half-heartedly trying to stuff it under the seat. I bent down to help him and stuffed it completely beneath, broke a nail, and then hit my head on his armrest as I stood up.

"Damn," I thought to myself.

"Thank you so much," said the yellow sweater.

"You're welcome," I replied, as I thought, *"Merry Christmas, Happy Hanukah, and kiss my ass,"* as I moved toward the aft section of the aircraft. *"Stewardesses indeed; the last time he flew must have been during the National Airlines or Pan Am era when they wore little hats , gloves , three-inch heels and had a weight requirement."*

Most of the other passengers seemed to know the drill, and other than reminding one or two passengers about fastening their seatbelts, raising their seatbacks, and pushing their bags completely under the seats, no more long explanations were necessary.

I opened my safety equipment bin and pulled out the life jacket, the oxygen mask, and the seat belt which I used for demonstration during the safety announcement, placed the items on the counter in the galley and checked to see if the paperwork was ready for the ramp guy.

"Bad news," said Scott the First Officer. "We have to go through de-icing. The weather is getting colder, and it's going to be a long night."

"Oh, no," I thought. So much for getting home after the Omaha round trip. We would have had to be right on schedule for me to have made the PFN flight home.

He handed me the paperwork, and I handed it out to the door to the ramp guy who was wearing a facemask and gloves and looked frozen. I closed the door and began the announcement, letting the passengers know that we would be slightly delayed due to de-icing. I didn't tell them that we were ninth in line. Quite frankly, I had only been through the de-icing procedure a few times, and it had been a while. While the passengers would not be pleased with the delay, I was happy to know everyone involved with the aircraft was on top of things because ice has been the cause of more crashes than the passengers would want to know about. I even had a habit of touching the top of the aircraft when I was flying in cold weather to see if ice was there as I entered; flight attendant training includes just enough information to terrify you of what can happen in all kinds of situations, most of which you have no control over.

I made the safety announcement, put away my equipment, and walked back through the aircraft, checking all of the overhead latches, seatbelts, and to make sure no one was in the lavatory. I then settled into my jump seat and smiled brightly at the passengers. I reached up for the phone in its cradle above my head as the two tones sounded meaning the cockpit was calling.

"We're still in line," Scott said glumly. "I am going to go ahead and make an announcement to the passengers to let them know what's happening, so get ready for the complaining to start."

"Thanks a lot," I laughed. "Merry Christmas."

There was no reply, and a few seconds later he began to describe the de-icing process and let the passengers know that we would be in line for several minutes before the chemical could be sprayed on the aircraft which would remove the ice and make us ready for flight.

Sighs of irritation and comments of dissatisfaction began immediately. I tried to feign a look of surprise as the announcement was made, as though the pilots had suddenly decided this process was a good idea all on their own . I put on my best, "this is not my fault so don't blame me look" as I continued to smile at the passengers.

Forty-five minutes later the cabin was an unhappy place to be; the holiday spirit had all but disappeared, when suddenly one of the passengers said , "Look at that, would you, they're on big ladders spraying the plane."

The poor guys in the freezing rain and sleet that was falling on Hartsfield Atlanta Airport were indeed on large scaffold-like ladders spraying in an eerie blue-green lighted area. After several minutes, we seemed to be finished and moved on, now moving toward our place in the long line of aircrafts waiting for clearance to take off. Five to six o'clock in the afternoon is an extremely busy time during the best weather conditions in the world's busiest airport, but today was just a mess. We seemed to be barely moving, then not moving, then barely moving again.

The passengers who were sitting in the front rows near my jump seat had plenty to say, and more questions than I could answer, but now those sitting in the middle and toward the aft section began to ring their flight attendant call buttons demanding that I get up to answer their questions. I stood up to move to Row 10 where an obviously irate, extremely large woman was looking at me expectantly, but before I could make it there, the familiar "ding-dong" two-tone call from the cockpit took precedence and I moved back to my seat and lifted the phone from the cradle on the wall.

"Hello," I said pleasantly.

"More bad news," said Scott. We have been in line so long that we have to go back and get de-iced again."

"You're kidding, right?"

"I wish I was; this is the trip from Hell. " He added, "Sorry, I know the passengers are gonna be pissed off."

"Not your fault; since we are going to be a while, could I serve them some snacks and drinks if I use the trays and not the cart?"

"Sure, why not. Wait a minute, Mike says to offer them comp drinks if you think that will help you out back there."

"Cool," I answered.

Alcohol always improved the general mood of the passengers because if they got a drink comped they felt like they were getting a little revenge on ASA . We weren't really supposed to announce comp drinks to the entire group, but rather, if someone asked for a drink we were to just tell them there would be no charge. I decided instead of just blatantly breaking the rule that I would compromise.

"Ladies and gentlemen, I regret to announce that we have now been in line for take-off so long that it is necessary for us to return to the de-icing area again; I apologize for this inconvenience, but of course your safety is always our first concern. "

"Unbelievable. Always sitting in damn Atlanta…that should be the motto of this crummy airline," remarked the gentleman in Row 2 who had not said a word until now. His face had been buried in the *Wall Street Journal,* and other than a sneer when he had to put his coat in the overhead instead of hanging it in a First Class cabin, he had been no problem.

"How do you become a flight attendant or pilot for ASA?" "Do you flunk out training for the bigger airlines?" he continued.

Some of the other passengers snickered at the comment, and there were plenty of worse things being said throughout the cabin.

I ignored him and continued the next part of my announcement as I glared at him for a moment, and then put back on my "friendly skies" face and said, "The bad news is we are delayed, but I will be offering a beverage service at this time , and the captain is pleased to offer our beer, wine, and cocktails in a complimentary service."

  A cheer went up. *"Fantastic,"* I thought. *"A four-dollar drink makes the world right again."*

"I will not be taking out the cart for safety reasons, but If you will have your order in mind when I come by, I will serve you using a tray a few rows at a time. Thank you again for your patience."

I went to the galley, took the tray out of the cabinet, grabbed a pen and a small notepad from my flight bag, and took orders from the first four rows. There are only fourteen rows on the regional jet, (actually only thirteen because they leave out the number thirteen for obvious reasons), and as a former bartender, I could serve really fast when I had to. I pulled out the liquor drawer and set two vodkas, a scotch, and a rum miniature on the galley counter, pulled open the soft drink drawer for ginger ale, a diet coke and a soda, and then opened the ice drawer.

"Son of a bitch!" I exclaimed before I could stop myself. "What in the hell do the lazy-ass food service people do ? How can you freaking forget to put ice on an airplane?"

I said all of this out loud and the first few rows didn't miss a thing.

I walked back to the telephone.

"Ladies and gentlemen, I have bad news and good news. The bad news is that the food service employees in Atlanta neglected to stock this aircraft with ice; the good news is that we can do shots."

I heard a loud roar of disbelief, then laughter, then stomping of feet as, "hell yeah's" began to echo through the rows of passengers.

"Ding-dong," I heard the familiar two-tone ring from the cockpit.

"Did I just hear you tell the passengers that WE can do shots?" asked Mike.

"Mind your own business, " I said as I promptly hung up the phone.

The holiday spirit returned, and the beverage service went much more quickly since I was pouring and serving juice and soft drinks with no ice and handing the adult drinkers a miniature and a plastic cup. The beer was cold anyway, and the wine drinkers were happy with whatever. Everyone had alcohol and pretzels and I sat down, called Mike to apologize for my abruptness, but found that he actually thought it was pretty funny. I reminded him that even though I had said WE can do shots, that unfortunately I wasn't doing one.

Once again we inched our way back to de-icing, and the passengers watched the process with a new sense of encouragement, fueled with the meager nourishment of the ASA galley. The Christmas spirit was alive and well again. I asked them if they would like to sing a Christmas carol to help pass the time.

"Yes !" was the loud answer I heard from what seemed like almost everyone aboard. I began to sing "Up on the Housetop," with everyone joining in and stomping their feet on the "Ho Ho Ho's" and clapping along.

"Ding dong." I picked up the phone again.

"What in the HELL are you doing back there?"

"We're singing Christmas carols," I answered innocently. "Look, you don't have any idea how awful this is back here. The two of you are up there with your coffee and cell phones and Popeye's Chicken, and I'm stuck with a shitload of unhappy people. Now leave me alone."

He hung up. We went through my entire repertoire of Christmas music, I served seconds to most of them, and I was cleaning up when the guys let me know that we were back in line for takeoff; we were seventh, and we were really going to Omaha.

Just like the *Wizard of Oz* I thought; they are really going home. I passed the information on to my weary passengers, and before long we were actually in the air, I made the 10,000 feet announcement, and the laptops and headsets came out.

"Ah, peace at last," I thought.

I moved over to the area directly behind the galley, out of the sight of the passengers, and sat on my bag in front of the bulkhead which separated me from the first row. I grabbed an apple out of my bag, made myself a hot chocolate, and stared out the small porthole in the galley. Now that we had climbed out of all the stuff, there was actually a beautiful moon. I rested for a few more minutes and then picked up the phone.

"Would you like some coffee or anything?"

"That would be fantastic," said Mike. "Put a shot of something in mine." When I didn't answer he laughed and said, "Just kidding. Two sugars would be fantastic."

"I would like hot chocolate if you have some," Scott requested.

I made their drinks, wrapped some Biscotti cookies in napkins and unlocked the door to the cockpit to hand their beverages and snacks to them.

"That was great work back there in Atlanta; the last flight attendant I had would have sat on her butt in the jump seat, stared at them and made them even madder. Thanks a lot."

"It was actually fun after we started singing," I said. "See you later. Call me if you need anything," I said as I closed the door and locked it again.

I made another pass through the cabin to check on everyone, pick up trash, and serve a few more refills; almost everyone was asleep now, and I returned to reading my Grisham novel.

Before I knew it we were on approach to Omaha, announcements had to be made, the cabin readied, and we were landing in more snow than I had seen since I was in high school in North Dakota..

"Oh no," I thought. " We are going to be stuck here all night."

 I was wrong. The passengers deplaned, a snow-plow was at work on the runway, and before I knew it, I had fifty new passengers, we were de-icing, making announcements, and back on the way to Atlanta.

The flight was completely uneventful; the passengers were tired from waiting on us to arrive, and most of them just went to sleep immediately. No complaints, no problems, just peace and quiet.

When we finally arrived back at Hartsfield, I was so tired I could barely drag my bags to the flight attendant lounge. All the departing flights from Concourse C had long since reached their destinations.

I found a reclining chair in the flight attendant lounge "dark room" where flight attendants slept when stuck in the airport, set the alarm clock on my cell phone, and curled up in a little ball under a blue Delta blanket.

" I wish I was at home," I thought . Then I fell into a deep sleep as I listened to the quiet, even breathing of other flight attendants in the lounge chairs and couches in the darkness.

Memoir of a Flight Attendant

## 5 EMERGENCY DOOR

"Today we are going to learn to open the emergency door and evacuation procedures," said David the flight attendant instructor.

"We will be going aboard the RJ (regional jet), and one at a time you will open the door, remove it, and place it on the emergency row seat . In a real emergency you would remove the door and throw it out the window in the direction of the engine after assuring the exit is clear of fire, smoke or water," he continued.

We had dressed casually today for our training , wearing khakis, oxford button-down shirts, and tennis shoes in the cool spring weather of the Atlanta Hartsfield Airport. I was still excited about my new career as a flight attendant, and I had studied the manual last night to make sure I knew how to open the door and prepare for the "emergency" evacuation.

We boarded the sleek Canadair regional jet and took our seats to wait our turn to demonstrate our emergency door-opening skills for the instructors. We were giggling about everything from the gay flight attendant, Russell, (who we adored) asking if he could wear the female uniform yesterday in class, to the customer service demonstration from the instructor who had shown us the "dip" motion used during serving (much like the Hugh Hefner Playboy bunny move) to assure we didn't put our butts in the faces of passengers as we were serving other passengers.

Suddenly we heard a scream from the middle of the aircraft where the over-wing emergency door is located.

"I did it! I did it" yelled Donna, the petite blonde in our class.

"Oh my God," yelled David. "Oh , fuck me," he continued.

"What happened?" "What's wrong?" the others in the class began to question.

"She threw the goddamn door on the tarmac," said Brandon, one of my classmates.

"A thirty-thousand dollar door," confirmed Susan the other flight attendant instructor.

Donna still didn't realize the mistake she had made.

"I did it!" she proclaimed triumphantly to the rest of us. "I did it," she said a little more uncertainly.

Her voice began to fall off in volume as she saw the looks of utter horror and dismay from the instructors and the rest of the class.

"Uh...come with me," David said to Donna, as he led her to the entrance door of the cabin.

Not only had she destroyed a thirty thousand dollar door, but she had put a thirty *million* dollar aircraft out of commission for the next thirty days…..and all of the ticket revenue it would have earned.

We never saw her again.

Memoir of a Flight Attendant

# 6 BROKEN URN

"It's not really broken, is it?" I exclaimed in disbelief.

" Yeah, whoever he *was* he's all over the floor in the galley now."

"Oh, my god," I said, "I told you to let her keep the urn under the seat."

"It was too damn big," Lashika said. "It wouldn't go all the way under the seat, and I don't care what they say, I don't like flying with dead people anyway."

I was horrified and also really angry with Lashika. She was Flight Attendant A on this flight, and a real bitch as far as I was concerned; the poor woman was on the next to the back row on an inside seat, and if Lashika had just kept her big fat mouth shut, she would still have the urn in her possession and we wouldn't be writing up paper work explaining why we were sweeping up someone's remains with a whisk broom and a dustpan.

I was already in a bad mood about this trip because I was the B flight attendant on the ATR; the A flight attendant makes the announcements and sits in the rear of the aircraft away from the passengers; the B flight attendant is like being a goldfish in a bowl; the seat literally folds down almost between the passengers on either side of the front row.

The woman in 15D had boarded the aircraft carrying what looked like a vase, and I didn't really pay that much attention to her, because it wasn't very large, and passengers bring all kinds of things with them---as long as they fit under the seat or in the overhead, it's not really a big deal.

I was doing the passenger count for the crew, making the usual checks for seatbelts and stowage of carry-ons, when I heard Lashika arguing with the lady in Row 15. Of all things, the vase was an urn containing the remains of her late husband, and besides being a little freaked out about it, Lashika was also being pretty unreasonable.

"If it don't fit under the seat, we gotta check it. I can put it back here in the baggage compartment behind the galley and you can pick it up on the way out," Lashika explained.

"But I don't want to do that," the woman argued meekly. " I am afraid it might get broken, and I can hold it here right next to the wall. It almost fits under the seat; it's just a little too tall."

"I said you can't do that; you either have to put it in the back, or I can call a gate agent and you can go back inside and work things out about getting it checked," she said curtly to the woman.

"Maybe we could bend the rules a little," I said. " It is a small container, and with the circumstances, I can understand why she would want to hold on to it ; it *is* almost under the seat…don't you think it would be ok?"

"No, I don't, and I don't appreciate you contradicting me. I know the FAA regulations and so do you," she said angrily.

"Fine," I said. I smiled sympathetically at the woman, but turned and headed for the cockpit with the passenger count as I watched Lashika take the urn from the woman, who by now had tears glistening in her eyes.

"What an absolute bitch, " I thought to myself. I knew Lashika was right about the regulations, but it was only an hour flight, and I am absolutely certain that someone ( even Lashika) must have broken a regulation or two along the lines working with ASA.

When I returned to the galley the urn was just sitting on the floor in the baggage compartment, the net was not in place nor the safety curtain which separated the baggage compartment from the galley, and Lashika was sullenly staring at it drinking a ginger ale.

"What's wrong? You aren't just going to leave it sitting there are you?"

"I ain't touching that thing again," she said. " If you want it someplace else you can do it. I don't like dead people."

"Oh for Christ's sake, it's just ashes," I said disgustedly as I tried to secure the urn. It had a lid on it that screwed down , and I made sure it was tight, and then placed it between two duffel bags, then pulled the safety belts, net and curtain. I hoped it would be ok.

"Your welcome, " I said as I turned my back and walked to the front of the aircraft, got out the safety equipment and did my

Vanna White demonstration; Lashika delivered the safety announcement in unrecognizable English, so fast that I hoped everyone knew what to do from watching me because they certainly didn't get it from listening to her.

The flight to Tri-Cities was unpleasant to say the least; it was a cloudy day and it was bumpy in the air; Lashika decided that we would just serve bottled water and snacks because it was "too dangerous" to take out the cart.

"I can serve and use a half cart; I don't mind," I said. The ATR carts have to be hooked together and are a pain in the ass to use, but if only one flight attendant is serving, he or she can use a half cart. Even though I was not happy that she wasn't willing to serve the passengers, I would rather do it myself than have them looking at me for an hour thinking how lazy I was for not offering them cocktails or soft drinks with their snacks.

" I am the A flight attendant, and I said it is too bumpy; we ain't taking out the cart."

"Fine," I retorted angrily. "But the pilots haven't told us to stay seated, and if we get in trouble I am telling them that you refused to serve or help me take out the cart so that I could serve," I fumed.

I stalked to the front of the cabin, folded down the jump seat, and immediately changed my expression back to the "fly the friendly skies happy face" as I sat down.

"So, is this going to be another one of those bread and water flights?"

The question came from the second row, aisle seat gentleman. He was not confrontational, and he actually had a pleasant expression on his face as he asked the question.

"I am afraid so," I smiled sympathetically. " I am not the senior flight attendant, and even though I disagreed with her, she has made the decision that the air is too bumpy for serving,"

I smiled to myself at the small, unprofessional transgression; I didn't mind throwing Lashika under the bus after the way she had behaved.

Some of the passengers were sleeping, others were reading , and many were staring at me with that belligerent look that plainly said, "where's the alcohol?" I have seen the look many times before. I know what it looks like.

Although the air *had* been bumpy for the first part of the flight, now the sun was shining brightly on the clouds beneath us, and to make matters worse, the First Officer decided to make a long, glowing announcement about the weather.

*"Good afternoon, ladies and gentlemen. We are happy to have you aboard this afternoon; we are flying at an altitude of 18,000 feet and looking at clear skies and a smooth ride all the way into Tri-Cities. Sit back and enjoy the flight, and thank you again for flying with us."*

"Great," I thought to myself. I smiled at the passengers, stood up, folded my seat, and walked calmly to the aft of the aircraft where Lashika was reading a book.

"Well, " I said.

"Well what?"

"How in the Hell do we not serve after that announcement?"

"Because we don't have enough time to serve and clean up before the initial approach announcement," she stated firmly.

"We have forty-five minutes," I argued.

"We are not serving, " she said as she continued reading, not looking up.

I continued to stare at her, when suddenly I noticed something protruding from the bottom of the curtain which separated the galley from the baggage area; I unhooked the curtain, and then I noticed several large pieces of broken pottery. I pulled back the safety net, and there, to my horror, was the urn, now in several pieces, with ashes scattered among the broken shards.

I yanked the curtain back , and tried to be calm as I asked, "Ok, so who's doing the paperwork ?"

Lashika still did not look up, and asked absently as she continued reading, "What paperwork?"

"The unusual occurrence." The unusual occurrence paperwork was a specialty item for ASA; when anything out of the ordinary occurred during the flight which could conceivably be a safety or public relations problem for the airline, an unusual occurrence form has to be filled out and signed by the Captain, then turned over to inflight supervisors.

"Why do we need to do all that? "It's bad weather, and we didn't serve," she replied.

"We have a bigger problem than that now," I answered. "Could you hand me the broom out of the utility closet, and a thank you bag."

Lashika finally looked up and started to say something, but then her eyes widened as she saw the mess all over the galley floor. She stood up, opened the utility closet and quietly handed me the broom, dust pan, and a grey plastic bag with handles which we call thank-you bags because we say "thank you" to passengers as we pick up their trash.

I began to carefully sweep the pieces of the urn and the ashes into the dustpan, trying not to lose any , holding my breath as dust rose from the bag. After several minutes of sweeping, emptying the dustpan into the bag, and holding my breath each time, I finished the task as Lashika stood watching quietly.

"So, now we have two unusual occurrence write-ups to finish before we get back to Atlanta."

"Two?"

"Not serving, and, oh yeah, we have a dead guy who's now in a dustpan and a thank-you bag instead of the urn he arrived to us in."

"Oh, shit. I didn't think we had to fill out an unusual occurrence for that."

"Well, I don't see how we can pretend it didn't happen. She *is* going to notice that her husband is no longer in an urn when you hand her a Delta trash bag."

Lashika moved to the galley and began rustling through the cabinet  looking for the forms she would need.

"You fill it out," she said as she handed the form to me.

I started to say something, but then thought, *"No, I want to describe the intricate details of this entire flight."* And I did.

I stuck the form in the little cubby in the galley to fill out during our return trip to Atlanta, and thought to myself, *"This will be a memorable moment for the flight attendant supervisors."*

The bells from the cockpit signaled initial approach, and I began checking the overhead compartments, picking up trash, reminding the passengers to bring their seats to upright position and to stow the items they had taken out as I made my way to my jump seat in the front of the aircraft.

Lashika, sitting in her jump seat in the back of the plane, was giving another totally unintelligible announcement with her reminder to shut down portable electronic devices the only understandable phrase.

Minutes later the ATR touched down in Tri-Cities, and as the passengers began standing to exit the aircraft, I could only imagine the level of sensitivity with which the urn incident was being handled as the remains of the poor woman's husband were returned to her.

I peered through the window near the front of the aircraft where I was seated, and I watched as the woman walked toward the terminal. Two gate agents were with her, and one of them was carrying the thank-you bag containing her husband's remains while another rolled her carry-on bag. She did not look back, and I sat down on my jump seat feeling sick .

"God, I hate this airline," I said out loud.

Margo Deal Anderson

Memoir of a Flight Attendant

# 7  CONCOURSE "A" JANITOR'S CLOSET

"What took you so long," Captain Mike asked Anita.

"The freakin' plane was late," she snapped.

Mike was a Captain who flew 757's for Delta, and Anita was an ASA flight attendant friend of Valerie who shared my crash pad in Atlanta with me and six other flight attendants.

She had met Mike on a flight to Aruba, and they met sporadically, believe it or not, in the utility closet on the A Concourse. Valerie laughed and said that Anita called it "getting her clock cleaned in the cleaning closet." On this particular evening, Valerie sat on the kitchen counter eating a protein bar, entertaining Elizabeth and I as she told us in great detail of her friend Anita's description of this afternoon's encounter.

"Mike fumbled with the zipper on his pants as Anita unbuttoned the navy blue vest and the coveted world scarf of senior ASA flight attendants," she said.

"Hurry," Anita whispered as she pulled her navy blue slacks down and hung them on the mop hook on the back of the door.

"Mike pressed against her as she hungrily kissed his open mouth and his hat fell onto the floor, " Valerie said dramatically as she took another bite of her protein bar and all of us giggled.

"You should start writing romance novels," Penny quipped.

"Shhh, let her finish," Ruth said, as Valerie continued.

"Oh, God," said Mike as he kissed her back while unfastening his belt and letting his pleated black pants fall to the floor .

"Oh, I missed you baby," she whispered as he pushed her against the wall , and pushed her lace panties aside with one practiced motion. He began to move his hips against her, using the wall as leverage.

"Oh, God," he whispered as her fingers grasped his hair and her legs wrapped around his waist.

"Where are you going this morning," Anita whispered in his ear.

"Aruba turnaround," Mike answered. "How about you?" he panted as he continued thrusting.

"Asheville, North Carolina , with eighteen hours at the Holiday Inn," she said.

"Are you close?" he whispered as he thrusted harder, pressing her breasts against him as he pushed into the wall .

"Yes," she gasped, and buried her head into his shoulder as he stopped moving and pulled her arms from around his neck.

The intimacy ended as quickly as it began. He reached for a paper towel from the cleaning supplies shelf, and then in a brisk, businesslike manner pulled his pants over his crisp white shirt, zipped his fly and turned away from her as she dressed.

"Was I too fast?" he asked.

"No," Anita lied as she buttoned first her blouse and then her vest, peering into the small mirror over the sink in the closet as she tied and straightened her ASA neck scarf.

"Maybe we can figure out who she is by the scarf," said Penny. "I want one of those map of the world scarves; the flight attendants who have those have been here the longest. "

"Forget it, Penny," said Ruth. "The only way to get one of those is if someone retires or dies and leaves it to you."

"Would you guys be quiet, and let Valerie finish," I said. Valerie smiled as she began once again to mimic Anita the flight attendant's breathy voice:

"Are you going anywhere this weekend?" Anita asked tentatively.

"Uh, I'm not sure…" hedged Mike.

"I have the next four days off after tomorrow; I'll be in about 10:30," she countered, trying not to sound desperate or needy.

"I'll call you," said Mike.

"Sure," answered Anita. She opened the door and walked out, headed for the ASA Flight Attendant Lounge. She did not look back.

"Oh, God, what a prince," I laughed.

"Of course; he's a pilot," Valerie said as she slid off the kitchen counter and started down the hallway to the room we shared. "Every flight attendant should have one."

# 8  STUCK IN THE HEAD

"Ding."

"Damn," I thought, as I looked up and saw the light indicating someone was calling from the aft head.  I was still in my jump seat and the seat belt light was still on.

"Idiot," I said to myself as I got up and started toward the head; we were still in a bit of turbulence and someone who got up before they were supposed to was apparently panicked in the john.  I held on to the back of the seats and made my way up the aisle to see if the obese woman from the aisle seat on Row 11 was ok.  I had already taken her a seatbelt extension and handed it to her discreetly before we took off, and the man next to her had bitched quietly under his breath about "why in the hell didn't she buy two seats?"

I felt sorry for her because I knew she was embarrassed by his comments, and I had glared at him as I handed her the seat belt extension and he shut up. She *was really large.* I guessed about 350 to 375 pounds and only 5'2" or 5'3". As I passed the asshole sitting next to her, he rolled his eyes at me and put the arm rest back down. She had moved it up when she initially sat down next to him, and in his defense, he had about six inches out of two seats to sit in.

I knocked on the door of the head and called out as quietly as possible, "Are you ok?"

"No, oh my God, no!"

"What's wrong? " I called out.

"I'm stuck on the toilet," she screamed.

"What?" I called back in disbelief.

"I'm stuck on the goddamn toilet," the woman screamed again.

The man in Row 11 went into hysterical laughter; the next three rows ahead of him turned around to see what was so funny.

"Shut up!" I yelled at him.

"Don't panic," I said to her through the door. "How are you stuck, like, between the walls, or actually stuck TO the toilet"

"Both," she said. "I pushed the flush button and now it feels like I'm just stuck; I tried to move and now my thighs feel stuck between the walls too. Oh my God, what am I going to do? What if it sucks me down?

"It won't," I yelled back through the door.

"Fat ass chance of that," laughed the asshole.

"You could try to be a little more sympathetic," I said.

"I can't," he answered. "I don't weigh 500 pounds."

I felt panicky. We didn't go over anything like this in flight attendant training, but I was positive she wasn't actually stuck to the toilet because of the flush mechanism. I remembered reading of an incident where a woman had actually gotten stuck to a toilet when the vacuum clean switch for the toilet was activated during flight somehow, and she was stuck on the toilet until the plane landed. There was no cleaning switch in the head on the regional jet as far as I knew; I felt that the woman was probably so large that she just couldn't stand back up in the very small space of the lavatory.

"Try to stay calm," I said to her. The two seats directly across from the head were empty, thank God. I sat down and asked her if she could open the door.

"I don't think so," she sobbed.

"Ok, just sit tight, and I'll be right back."

*"Sit tight, now why in the hell did I say that?"* I thought glumly.

I called the cockpit.

"Hey, Margo, everything all right? How about a cranberry juice?"

"Not yet, I have a problem."

"Ok, what's up?"

"The heavyset lady you were laughing about earlier is stuck in the head."

"Say again?"

"She's stuck in the head. "

I now heard *two* assholes in the cockpit laughing hysterically.

*"Great,"* I thought, *"more comedians."*

"Look you guys, this is not funny, I need help."

I took a deep breath, opened the overhead compartment and pulled out my small bag with my flight attendant manual, pens, and notebook in it. One of the easiest things they taught us to do in training is to open the door to the lavatory if someone can't open it form the inside. You can simply insert the end of a pen in the hole and slide the latch open. I headed for the aft of the plane where half of the plane was now openly laughing and watching to see what I would do next.

I knocked on the door of the head.

"She's still there," laughed the man sitting in the seat where she had been sitting.

"Not funny, " I said to him.

"Are you alright, ma'am?" I asked tentatively. No answer.

"Ma'am, are you alright?" I repeated.

"This is so humiliating. What are you going to do?" The voice sounded very small. She was really crying now.

'First, I am going to open the door, then we'll figure out something; don't worry."

"Noooo!" she shrieked. " Everyone will see me."

More laughter from the passengers.

I looked around desperately, and I saw a woman who was *not* laughing sitting in Row 9. I knelt down beside her seat and asked her if she would assist me.

"Could you hold this blanket up for me?" I asked her. Row 14 was empty, thank God, and it was the row directly across from the head.

The lady held the blanket up to shield the last row from view as I opened the door.

"*Oh, my God*," I thought to myself; "*this is unbelievable.*"

"Hi, " I said to the woman, who was basically stuck between the two walls of the tiny lavatory. Her legs looked like two gigantic sausages.

"Ok. Let's see if I can help you out of here." I crossed my hands, put a foot on each side of the door and pulled as hard as I could. Her hands were sweaty, and they slipped out of my hands. I fell backward into Row 14 and hit my head on the arm rest.

"Ouch," I said.

"It's useless," the lady whined.

"What's your name?" I inquired.

"Tiffany," she said. The name sounded so delicate and airy.

"Ok, Tiffany. Just relax, take a deep breath, and we'll try again."

"What's your name?" I asked the lady with the blanket.

"Martha," she answered.

"Thanks, Martha. If you will just keep holding the blanket in place I'm going to try again. "

Just then I heard the familiar chime of the flight attendant call button. I stuck my head underneath the blanket to see what was wrong now.

"Hey, could I get a drink, please, or is that too much to ask?" yelled the guy in Row 5.

I ignored him and ducked back under the blanket, turning my attention to Tiffany.

"Ok, Tiffany. This is going to be personal, but I don't know anything else to do." "I am going to reach behind you and pull you toward me a little. Can you help me and lean forward?"

She nodded miserably and leaned forward.

"She's never going to be able to pull that hog out of there," the man in front of us commented.

"Shut up," I said over my shoulder.

Tiffany said nothing but tears were rolling down her face and she was sweating profusely.

My hair had come loose from the French twist and was hanging in my eyes as I took both of her hands, braced my feet against either side of the door of the lavatory and took a deep breath. I am

5'10" and weight about 145 pounds, but I was no match for Tiffany.

"Ok, Tiff, we're gonna give it another try. On the count of three you try to stand up, and I am going to pull you out of there."

"One, two, three, go!" She leaned toward me, slowly, and I pulled with all my strength. Just as I thought my arms were going to come out of the sockets, she began to stand and I fell backward as she fell out of the head on top of me in the back row, partially in the aisle.

Martha dropped the blanket over her, and now we were all stuck between the head and Row 14.

"Great," I thought. I scrambled to get upright, stretched one of my legs over Tiffany, and vaulted into the aisle ahead of her. The blanket wasn't hiding much.

I held it up for her while she struggled to pull up her elasticized pants.

"I'm moving before she gets back in this seat," said the asshole.

"Take Row 14" I said to him. "It's empty."

"I would if I could get past her to get there," he quipped.

"Walk forward a few rows," I snapped. " Let her sit down and then you can move."

"I'm writing a complaint to Leo," he said.

"Go for it, " I answered.  Leo was the President and CEO of Delta Airlines, and I really doubted this "platinum medallion" prick knew him personally, but then, you never knew.

I helped Tiffany into her seat, moved out of the way of the "prick" so that he could get into Row 14, and then I headed for the galley in the front of the plane.

I was exhausted, and I glared at the passengers as I brushed a wisp of hair out of my face, turning away from them to get a bottle of water for Tiffany.

 "Good Lord,"  I thought, what else can possibly happen today?

"Ding-dong."  Cockpit calling.

"*Really*?" I thought.

"Yes?" I said pleasantly into the phone.

"So, how's it going with the fat lady?" asked the First Officer.

I hung up.

"Ding- dong"

"What?" I answered.

"Ok, I'm sorry  Are you ok?"

"Yes, " I answered.  "No thanks to you two."

"Can we please have a cranberry orange juice and a coffee black".

"Yes," I said curtly.   "Give me just a minute."

I started a fresh pot of coffee and began mixing a can of orange juice and cranberry juice in a cup  for the First Officer with just a little ice---I tried to keep my crew happy, but today my heart was not in it.

"*Assholes*," I thought to myself.

## 9 KEVIN DAVIS

I had five minutes before duty-in, and I ran into the flight attendant lounge bathroom on the way. Four or five flight attendants were gathered around one of the stalls laughing, and I turned to go in one of the stalls on the other end.

"Margo, come look at this," motioned Christa, one of my friends who flew with me sometimes.

"I only have a couple of minutes before duty-in." I waved her away and started into the stall.

"No, don't use that one, use this one," she giggled as she pulled me toward the stall in front of the group of flight attendants.

"Why, what's up?" I asked suspiciously.

She practically pushed me inside, and seeing nothing wrong, I locked the door and sat down.

Still wondering what the big joke was, I looked at the back of the stall door while sitting on the toilet.  Someone had scrawled in large, black letters:

## KEVIN DAVIS HAS A BIG D---- AND LOVES TO LICK P-----!

"Oh, my God," I laughed, as I read.  "Who's Kevin Davis?"

"That's what we want to know," Christa laughed.

"Is he a flight attendant?" I asked.

"Are you kidding; if he was this would be written in the men's room," laughed Marion.

"All male flight attendants aren't gay," I said.

"Which ones do you know who are not?" Christa challenged.

I began to think of my male flight attendant friends; I honestly could not think of one who wasn't gay.

"Well, I'm sure there are several; I just can't think of any at the moment."

"Exactly," quipped Christa.

Annoyed, I flushed the toilet, fumbled with my panty-hose which were already sweaty in the Atlanta humidity, just from the short walk from the employee parking lot , to the bus, and inside to the flight attendant lounge.  I pulled them up, even more annoyed with the company policy requiring female flight attendants to wear pantyhose, even in the sweltering heat of the Atlanta summers.  I quickly opened the door, moved to the sink and began washing my hands.

"Someone probably just made that name up for a joke, and it was good for a laugh. I've got to go and duty-in. Have you started a trip yet?"

"I'm headed for White Plains," sighed Christa. "I hate that place, I hate my crew, and I hate flying by myself on that wretched long flight with fifty whining passengers."

"Sorry, " I replied as I pushed the door open and hurried in the direction of the duty-in station. I was not in the mood for listening to complaints from anyone today; I had a four-day trip facing me on the ATR, and I had promised myself to remain in a happy mood regardless of who I was flying with. As I reached the window and repeated my name a second time for the duty-in guy, I heard someone in the hallway asking about Kevin Davis, and I giggled to myself.

"What's so funny," asked the flight attendant standing behind me.

"Nothing," I said. "I was just thinking of something that happened this morning," and I quickly pulled my bag behind me, headed for the lounge and the door which would lead outside to the tarmac.

 A blast of hot , sandy air hit me in the face as I left the air-conditioned flight attendant lounge, and I wished for the ten-thousandth time that I was a Delta flight attendant, walking onto a jet way to enter a large 757 instead of pulling my bag across the black, sticky asphalt of the tarmac headed toward a stairway leading to the most ridiculous looking aircraft in the Delta Connection fleet, the ATR.

"Hey, green eyes," shouted my favorite ramp guy, a tall, lanky African-American guy named Stefan. He was always nice to me, always helped me carry my bag up the stairway, and always had a beautiful smile, the whiteness of his teeth in sharp contrast to his dark, mahogany skin.

"Hi, yourself," I smiled to him as I checked the number on the aircraft and started moving toward the stairs.

"You going to Panama City today?"

"As ridiculous as it sounds, I just commuted from there, and now I'm headed back." I would bid the Panama City trips whenever I could because it meant I could spend the night at home and be with Lee. I would actually make one round trip to Panama City, a round trip to Tri-Cities, and then an overnight in Panama City before the day was over. I was always amused with passengers who asked me if I flew only back and forth from Panama City all day long.

"I love going to Panama City--- all those beautiful girls in bikinis. Every time I get days off, I go there if there is a seat. "

"Where do you stay? On the beach?"

"No, I stay with my girlfriend down there."

"I thought you had a girlfriend here in Atlanta?

"I do," he grinned at me as he said it.

"You are bad," I smiled as I began climbing the stairs, dragging my bag and my lunch cooler.

"I'll get that for you," he said as he took the bag from me.

"Thanks," I smiled gratefully. "I will get you a ginger ale."

"Just a water would be great," he said. "You got a reserve flight attendant flying with you today. I just met her, and she cute."

"Ok." Stefan thought all women were cute. I put my lunch in the overhead over the first row of seats, and reached down to get my bag as Stefan handed it up to me. I pushed it into the overhead as well, and then grabbed two waters out of the drawer and gave them to Stefan. I noticed the other flight attendant was coming out of the door at the front of the aircraft which was where the "B" or newer flight attendant on the ATR would sit. I waved at her.

"Thanks." Stefan smiled at me, shook his head and skipped the last two stairs as he ran to help unload bags from the ATR parked nearby. I began my safety check, and noticed that the First Officer was walking around the outside of the aircraft doing his checklist for the "walk-around," looking for anything unusual on the outside of the aircraft. The ATR is backward in every way, even with the entrance at the back of the aircraft. The lavatory is the first thing passengers see besides the galley, and they are always amazed to be entering from the back, in addition to also being on a "puddle jumper" as they love to call the sixty-six seat aircraft.

"Hi," I said to the other flight attendant. I'm Margo."

"Ginny," she said. It's very nice to meet you. I am on reserve. How long have you been flying?"

"Three years." She was asking because she was hoping she had been here longer than me and could sit in the back and be the "A" flight attendant. Her eyes fell in disappointment, but I did not feel guilty. I had been the "B" flight attendant on this wretched airplane more times than I could count. I smiled at her and tried not to look like the cat who had just swallowed a canary.

"Do you know Kevin Davis?" she asked innocently.

"Oh my God, you must have been in the bathroom in the flight attendant lounge," I laughed.

"I was," she giggled. "Who do you think he is?"

"The product of someone's imagination who is trying to get some new gossip started," I said matter-of-factly. Playing jokes was part of the flight crew world.

"I don't think so," Ginny said. "Everyone in the flight attendant lounge was talking about him and trying to figure out who he is. He must be a really ripped Captain with six packs or something."

"If he exists it won't take long," I laughed as I went back toward the galley in the aft.

I finished my safety check, made sure there was ice in the ice drawer, liquor in the liquor drawer, and dry ice on the beer and wine. When I was satisfied that I would not face a mutiny from the vacationers who always wanted to drink on the way to my hometown , known affectionately as the Redneck Riviera, I put ice in a cup and poured a Diet Coke for myself. I watched Stefan pull a baggage cart next to the entrance to the stairs, and he yelled up at me, "here they come."

"Passengers are coming," I called to Ginny who was standing in the hallway of the ATR, a peculiar area leading to the cockpit, with baggage on either side. She was talking to the Captain, but turned and walked toward me to help with greeting and getting the passengers settled. I had not spoken with the crew yet, but it would have to wait.

I saw the passengers walking onto the tarmac, holding their hair, their skirts, and squinting into the sun as they appraised the long walk from the door of the terminal to the aircraft and began following the covered walkway with directions from the gate agent. I watched with amusement as the first passenger began to argue with Stefan about "valet bag check" which was always a problem. Passengers who were frequent-fliers with Delta did not understand that the overhead compartments on the Delta Connection aircrafts were smaller than the MD88's, 737, 757, and other aircraft of the Delta fleet. They simply could not believe that their carry-on luggage could not be carried on. I watched Stefan shake his head in disgust as the first woman pulled her oversized carry-on right past him, insisting that it would fit. She began dragging it up the stairs as I stood at the top .

"Ma'am, your bag is not going to fit in the overhead," I said apologetically.

"Yes it will," she shouted as she struggled with the bag.

"If you will just leave it at the bottom with the pink tag on it, you will get it back as you leave the aircraft in Panama City."

"No I won't," she snapped. "Every time I check a bag you guys lose it. I'm not checking it."

"We are only going to Panama City," I said patiently. "You can watch him put it on the plane. There is no way for it to get lost."

She ignored me and entered the door pulling the bag behind her. As she reached her seat at Row 7, she hoisted the bag above her head and began trying to shove it into the overhead, as I turned my attention to other passengers as they were entering.

"Good morning," I said. "Panama City?" Part of the greeting on the Delta Connection flights always includes mentioning the destination city because so many flights leave from gates which are in close proximity that it is not uncommon for passengers to board the wrong aircraft.

"Yes," said the tall gentlemen. "Row 4?"

"Almost all the way to the front of the aircraft, sir," I said as I pointed.

"Good morning," I repeated as the passengers continued boarding, a mixture of families with children, business travelers, and college and high school graduates heading for a week of partying on the "World's Most Beautiful Beaches."

I noticed the passengers were not moving very well down the aisle, and as I looked over their heads, I saw Ginny arguing with the over-sized carry on lady.

"It's not going to fit," Ginny shouted. "You have to valet-check it."

"Ok, " the woman shouted, "you can carry it back down those damn stairs. It's not my fault the overhead compartment is so small. This is ridiculous."

I quickly made my way through the passengers and grabbed the handle to the bag. "I've got it," I said. "Don't worry ma'am; your bag will be fine." I turned my back to her and began making my way back to the doorway with the heavy bag. I placed the bag behind me in the galley and continued to greet the passengers. Ginny made her way to the galley and got the passenger count form to begin counting the passengers ; the " B" flight attendant was in charge of the passenger count for the pilots and the safety demo.  About half the passengers were aboard, and I noticed Stefan was once again explaining to another passenger about the pink valet tag on carryon bags.  This time the passenger was a man wearing a business suit and he was hanging on to the handle of the rolling bag as though it contained a million dollars as he argued passionately with Stefan, who continued to smile and nod his head.  As the man began climbing the stairs with the bag, he was breathing hard and his face turned red.  I stood at the top of the stairs and could see Stefan rolling his eyes as the man made it to the top and planted the bag at my feet.

"I am a Platinum Medallion flier, and I cannot believe I can't get some help here," he gasped as he stood in the doorway, holding his jacket over one arm revealing wet sweaty splotches under both arms of his light blue shirt.

"I'm sorry sir, but I am cannot step out of the aircraft to assist you according to company policy. "

"I'm in Row 12," he said, mostly to himself as he began pushing the bag ahead of him toward his seat.  He laid his jacket on the back of the Row 12 aisle seat, lifted his bag, and began trying to push it into the overhead.  He turned it around and tried again.  Still no luck.  He threw it to the aisle, sat down, and rang his flight attendant call button.  I ignored him and continued greeting passengers.  I watched out of the corner of my eye as Ginny went to the man, argued with him for a moment, and then began dragging his bag toward the galley.  She was a petite brunette , and the neat French twist with which she had started the flight had already began falling, leaving sweaty tendrils across her forehead.

"Row 12, Seat B is a prick," Ginny gasped as she pushed the heavy bag behind me into the galley and walked back to continue working on her passenger count.

"A Platinum Medallion prick," I said to myself. I had some pink tags in the galley, and I placed a pink tag on each of the bags in the galley, shoved them into the cargo bin behind me, and began fastening the safety net and curtain across the cargo area. One of the positive attributes of the ATR was the accessible cargo area behind the galley. I sometimes placed my own bag back there, and in the days before 9/11, commuting crew members would hide back there and sleep on top of the bags as they flew home.

Finally, all of the passengers were aboard, and as Ginny came to the galley to get a bottle of water, I went forward to meet the Captain and First Officer. I knew the Captain, a small, pleasant guy whose name was Adam. I had flown with him many times, and I really liked him. Adam had flown helicopters in the Army, and in addition to being a well-respected pilot, he had an incredible sense of humor. I didn't know the First Officer, who introduced himself as David and immediately asked for a cup of black coffee.

"Sure," I said. "I'll be right back with it.

"Hi Adam," I said happily, as I left the cockpit and headed for the galley. This trip would be fine no matter what. If the Captain was not an asshole, a four-day trip could be a good thing, and Adam always looked out for his flight attendants , unlike some of the Captains I had had the misfortune of being stuck with time after time.

I noticed a woman with her back turned to me as I approached the galley; she had the flight attendant jump seat folded down and was bending over it. A wave of stench invaded my nostrils as I got closer, and then I saw that she had a toddler stretched across the jump seat changing his diaper, and she had already placed the diaper and its contents on the counter of the galley.

*"You have got to be kidding me,"* I thought. *"I have seen it all now, a shitty diaper lying on the galley by my coffee pot."*

I cleared my throat, grabbed a Delta thank-you bag out of the utility closet, and said in the most even tone I could manage, "Excuse me, ma'am, could you please take the diaper off the galley counter and put it in this bag? "

She turned with one hand holding the squirming little boy with everything exposed , grabbed the diaper, deposited it in the thank-you bag, and then turned back to the task at hand. She finally got the diaper on the forty-pound little boy, grabbed him by the hand and headed back to her seat, leaving me holding the bag.

"Thanks," I said, under my breath. I waved at Stefan, who was just getting ready to roll the cart with the valet-checked carry-on bags to the baggage compartment, and with my sweetest smile, I beckoned him to the stairs. He ran back to the stairs and bounded halfway to the top where I handed him the thank you bag and said, "could you please be a sweetie and throw this away out there----it's a dirty diaper."

"Shit," he said as he looked into the bag. "Shit," he repeated, as he held onto the bag and ran back to the cart. He waved goodbye without looking back. I took some hand-sanitizing towels out of their foil packets and wiped down the galley counter, went in the lav and washed my hands with soap, then made the coffee for the crew. The First Officer had asked for black, and I knew Adam liked his the same way.  I asked Ginny to take the coffee up while I got ready to secure the galley, close the door and make the safety announcements, and she walked back up the aisle.

"Guess what they were talking about when I took them the coffee," she giggled.

"I have no idea," I said.  "What?"

"David was asking Adam if he knows a guy named Kevin Davis."

I handed the paperwork out the door to Stefan,  who apparently was still mad at me about the diaper because he didn't give me the usual flirtatious smile; I lifted the stairs, secured the door, and sat down to begin the long, arduous memorized announcement for ASA flights as Ginny made her way to the front to hold up the demo seat belt , oxygen mask and life preserver while I spoke.  I could say the entire safety announcement in my sleep, but every time I recited it, I thought bitterly of the Delta flight attendants who either turned on a video safety announcement or read from the flight attendant manuals.  ASA is the most old-fashioned, behind-the-times airline.

" *I don't care if I have to go on reserve again, I am going to apply to Delta as soon as this trip is over*. "   I promised this to myself every time I started a four-day on the ATR, but as soon as it was over I would start thinking about being on reserve, going through training again, and Delta Connection didn't look so bad, especially since we had the same flight benefits as Delta employees now.  I also wasn't sure I wanted to go through finding a crash pad and living in Newark for an indefinite length of time.  Lee and I were in a very good place with our relationship now, and moving to Newark might not be the best decision at this point in time I would tell myself.

   Ginny put away the demo items, and I watched as she walked up the aisle toward the galley where I was sitting, checking the overhead compartments as she walked to make sure they were closed.  We were in line for takeoff, and she stood in the galley for a moment out of sight of the passengers.  I felt a moment of sympathy for her because the "B" flight attendant had to sit facing the passengers with her knees almost bumping the passengers on both sides from Row 1.  I had actually had passengers put their feet on the sides of my jump seat before, leaving footprints on my navy blue slacks or skirt.

   "Where's my coke?" she inquired as she looked on the galley counter with a puzzled expression.

   "Sorry, I trashed it after the lady put the dirty diaper almost on top of it."

   "What!" she exclaimed.

   "She actually changed the kid's diaper in my jump seat and then threw the diaper on the counter."  "I'll make you another Coke as soon as we take off. "

   "Thanks." She was still shaking her head as she went to her seat and fastened her seat belt.

     I fastened my seat belt, said my quick prayer that I always say in the moments before takeoff, and then felt the familiar acceleration as the ATR gained altitude, turned and headed  south for Panama City.  I picked up the copy of the new monthly ASA magazine which had just been placed in the seat pockets and began to flip through it.  Suddenly my eyes were drawn to the photo and caption on the cover. The employee of the month was a thin, partially-bald gate agent who could have been a star in the old *Revenge of the Nerds* movie. I began to laugh uncontrollably as I read the caption on the cover photo, *Kevin Davis is ASA Employee of the Month.*

## 10  CRASH PAD

A luxury apartment in Stockbridge.  Three bedrooms, two baths, a gym, and a pool.  Eight flight attendants sharing three bedrooms? I could see no way this would work, but the flight attendant supervisors assured us we would never see each other, and we would have plenty of room.

"Why do they call it a crash pad," asked Penny.  "What a negative term to use for flight attendants and pilots," she reflected as she picked at an imaginary flaw in her perfect manicure.  Penny was a doctor's wife who was recently divorced and had decided to try her wings in the employment world.  She had appointed herself to our group which had haphazardly formed as we huddled together miserably during the lunch breaks during training.

"Because you only crash there for a few hours to sleep in between trips," answered Valerie matter-of-factly.  "It wouldn't matter if there were twenty people staying there; you never see each other, and when you do you are too tired to care who's sleeping there."  Valerie had a friend who dated a Delta pilot, and she was a never-ending source of information for us, both good and bad.

At this point in time we were all still living at the Embassy Suites hotel for training, but graduation was only four weeks away-- for those of us who still felt we had a chance of graduating. I am a planner; I have to know where I am going to be living more than twenty-four hours in advance, and we had already been told that we should be prepared to be called to fly beginning at midnight on the day of graduation. Our time as reserve flight attendants would begin, and while on reserve you have to wear a beeper. If crew scheduling calls you to fly, you have one hour to be at the airport or you could lose your position as a flight attendant almost before you got started.

We did not want to pay a great deal of money for a place where we would be spending so little time, but on the other hand, we didn't want to live in a dump either. After much discussion we finally agreed on the apartment in Stockbridge, and Valerie and I agreed to sign the lease since no one else was willing.

" Great," I thought. I now have a one year lease in Atlanta for a three-bedroom apartment I can't afford, and I have not yet passed the flight attendant training.

Valerie and I claimed one room, Penny and Rachel took another, Ruth and Belinda took the third, and Mary and Elizabeth were left with the living room. Valerie and I had demanded the bedroom with the second bathroom because of signing the lease; we were entitled to at least one perk; we did agree that others could use the bathroom when we were not there, however.

After training was completed we would be given one weekend to move to Atlanta before graduation, and I started making a list of what I would be bringing from Panama City. Because I wanted to commute, meaning I would live in Panama City and fly or drive to

Atlanta for trips, I was determined not to move too much stuff.  I remembered the queen-sized inflatable mattress stored in the top of the closet at home.  All I really needed was the mattress, some sheets and towels, maybe a small lamp, and a few dishes, pots and pans, and cooking utensils.  The crash pad was going to be very temporary for me; I would not admit it, but I was extremely homesick, I missed Lee, and I was sorry I had ever thought of being a flight attendant.  At this point, I didn't have a job, or a commitment from my boyfriend, and my pride would not let me admit to anyone that I had probably made a huge mistake.  As I had done most of my life, I listened to the voice in my head which belonged to my grandmother  as well as my mother saying, "you made the bed, now you have to lie in it."  I'm not sure why I felt this was a truism to live by, but the things from childhood somehow seem to stick.

"I hope the rest of them pay their rent and their part of the electricity, cable and water," sighed Valerie.

"Me too," I said glumly.  "At least we have that taken care of.  I mean, half of our class is consumed with the search for crash pads.  All we have to do now is get through four more weeks of training and graduation."

We stopped at the mall on the way back to Embassy Suites and purchased six month contracts for beepers which would soon become objects of great disdain. At that time, beepers were the thing because smart phones had not yet been invented.  We then went to happy hour at Embassy Suites and began drinking wine as though we did not have a test over the parts of the ATR aircraft tomorrow.  Our class was down to about thirty now; we had started with over fifty. Almost all of those who remained were far ahead of us in the two-for-one happy hour.

"Cheers," Valerie said as she lifted her glass of wine.

"Here's to the friendly skies," I replied, wondering when in the hell I would learn the parts to the ATR before 7:30 a.m. tomorrow. Making less than 90% on the written test given each day was a ticket home; I wasn't ready for that humiliation. I gulped down the wine and headed upstairs to my hotel room to study the parts of the ugly, turbo-prop aircraft known as the ATR; it looked like an aerodynamic mistake.

## 11  DELTA AND ASA

I could not believe I was sitting next to Karl Rove on the last leg of our flight from Costa Rica which was from Atlanta to Panama City, Florida.  I am really an Independent voter, and I was not a huge fan of Mr. Rove, but I must admit, we had some really interesting conversation on the flight and I was surprised at how personable and interesting he was; he actually has a home in Rosemary Beach, an upscale gated community about 25 miles west of Panama City.

I did not ask, but assumed he was vacationing with his wife, who was also on the flight.  We talked for about fifteen minutes on topics including my singing career, President Bush, the war in Iraq, and how much tequila I had consumed in Costa Rica.  My husband Lee and I were returning home after three weeks in Quepos , Costa Rica where I had been singing at the Kamuk Casino and Wacky Wanda's Bar Central.  My long blonde hair was wet and in a French braid, and I had on jeans with holes in both knees.  I faintly remember telling him how I had sung the National

Anthem for the Indianapolis and Louden, New Hampshire NASCAR races, and how I was sure he could get me a gig at the White House. To his credit, he did give me the email of his personal assistant, and later when I contacted her, she actually replied. As a side note, after almost eight years with President Obama, I am actually on the same side with Karl now, and I watch with great interest when he brings out his white board on FOX News. After a few minutes of polite conversation, I asked for a photo opportunity with Mr. Rove, to which he politely agreed, and then I left him alone.

After sleeping for the last two hours while my husband Lee watched the inflight movie on our flight from Costa Rica, I was awakened by some turbulence as we began to make our approach into the Atlanta Hartsfield Airport. In addition to our new friendship with Karl Rove, Lee and I had been also accompanying Bobby, the husband of Wacky Wanda, owner of the bar, on his flight home. Bobby, bears a remarkable resemblance to Imus (of Imus in the Morning), right down to the straw cowboy hat, and this particular flight was just two weeks after the big media coverage of Imus' nasty comment about an African-American girls' basketball team. My husband found particular joy in telling everyone that Bobby was Imus, and unfortunately for Bobby they believed this lie. Bobby is a very young sixty-eight year-old man, and he is also a fan of fine wine and beautiful women. The flight attendants love him because he compliments, tips, and believes that he is still flying in the 1960's on Pan Am in happier times.

When we finally arrived in the baggage claim area of Atlanta customs, I was holding Bobby on my shoulder to prevent him from falling, and when I asked him to describe his luggage, he replied, "It's black." When I asked him how many bags he had checked, he replied, "I don't know." At this point, I was looking in the large checked items section for my amplifier, speakers, guitars, surfboard, and a random number of black bags, along with a surely illegal amount of checked Nicaraguan Flor de Cana rum which belonged to Bobby. To my great surprise we made it through customs, and somehow stumbled to Concourse C to wait for our connecting flight to Panama City, Florida, the last leg of our trip where we met Karl Rove, and waited together for what turned into several hours.

    By this time, I was at the end of my flying career with ASA, and I was flying as a part-time flight attendant based in Dallas, Texas. Lee and I were flying more hours as non-revenue passengers than I was flying as a working flight attendant. My days with ASA/Delta Connection were numbered and I knew it, but we were having a fabulous time flying everywhere. I had shared with Lee all of my stories about how ASA maneuvers flights to their own advantage with little or no thought of the impact on passengers; gate changes, cancellations, rude gate agents ,and delayed flights were all part of the low expectations which the regular passengers of ASA at that time just accepted as routine. This day was to be no different. We had arrived in Atlanta in the early afternoon, and our gate had already been changed twice. Lee was mocking the gate agent with his best , exaggerated Montgomery , Alabama southern drawl saying in a loud voice for all in the gate area to hear, " Attention everyone; I know what the last question on the test is to become a gate agent in Atlanta: " I sure am sorry for the delay here in Atlanta!" The other passengers roared with laughter, and began to imitate his "sure am sorry" speech as they walked to the next gate.

    "Being sarcastic to the gate agent is not going to get us home,"

I reminded him angrily. His response to this was to begin telling the passengers that they could probably expect another gate change because the Panama City flight was not really in maintenance but had been sent on a round trip to Chattanooga and would return in about an hour at another gate. Whether this was the case or not, approximately twenty minutes after he made this comment, the gate agent announced that we would be changing gates, *for the third time.*   As the herd of passengers moved out of their seats toward the new gate at the other end of the concourse, Lee began making  inflammatory gate agent voice imitations, and to my horror, the other passengers began to imitate him even more aggressively. I would like to mention that I did *not see Mr. Rove taking part in any of these shenanigans.*  He sat quietly reading a book and followed the herd from gate to gate as necessary.

   We had now been waiting for several hours for the connecting flight to Panama City, and the comments from my husband and our fellow passengers had run the gamut from "we could have rented cars and driven by now," to "ASA...always sitting in Atlanta," to "ASA….always sitting on their asses." This group of passengers was ready for a riot, and my husband was ready to lead them. Bobby had been asleep, but for some reason , he suddenly awakened and pulled himself to a standing position. My husband noticed immediately, and announced to everyone that Imus in the Morning was among us. Passengers surrounded Bobby asking for autographs, and I quickly ducked into the nearest ladies room, praying that boarding would commence immediately.

   Thankfully, we did board our flight soon after this, and in spite of bad behavior, everyone was allowed to board and continue on to Panama City, Florida. Although loud, obnoxious comments and rude behavior abounded among the passengers who waited almost nine hours for a flight which would last only 55 minutes, I removed myself both physically and mentally from the situation by sitting a good distance from all of them in the Atlanta airport. I was still an employee, and we were flying on non-revenue tickets; even though I knew my husband and the others were justified in their declarations, I was hoping not to lose my flight privileges just yet. Looking back, I realize how many times I had been embarrassed that I was a flight attendant with ASA. When anyone asked me who I worked for, I always said Delta, never Delta Connection or ASA.

Margo Deal Anderson

## `12 Brown Sugar

The first officer could not take his eyes off Janelle as she boarded the ATR. I was arranging the galley and getting ready to do the safety equipment checks when she entered with a big smile.

"Hi, Margo."

"Hi, Janelle"

"God, it's hot, " she said as she opened the overhead and stashed her small crew bag inside, pushed her lunch bag beside it, and grabbed a napkin off the galley counter to wipe the perspiration from her face.

Paul, the First Officer, was still staring, and had forgotten to finish pouring his Mr. Pibb over the cup of ice I had just handed him . He was taking the place of the First Officer who had been flying with us on the first two days of the trip, and he had not met Janelle yet.

"Hi, I'm Janelle," she said as she stuck out her hand and flashed her beautiful smile again.

"Paul," he said. "Nice to meet you." He quickly turned and walked up the aisle toward the cockpit.

Janelle was an absolutely drop-dead gorgeous African-American flight attendant. She had an almost bronzed glow about her, bluish-grey eyes, and a figure that would have made the cover for Sports Illustrated swimsuit edition. She was one of the few flight attendants who wore the navy blue dress which was one of our uniform choices, and she had obviously had it tailored to fit her, because most of the dresses had an A-line cut which gave no idea of whether or not the flight attendant had a figure or not.

What I had noticed that was different about Janelle today was that she had stared just as much at the First Officer as he had at her; she was usually oblivious to the stares she got, or just ignored them.

"Who is he"? she asked. "I've never flown with him before."

"I don't know," I answered," this is the first time I've ever flown with him as well."

"I'm going to see if Jonathan wants anything to drink," she said.

"He's not up there yet," I said.

"I know," she answered with a wink , and headed toward the front .

I laughed to myself and continued stocking the galley, thinking what a nice person she was. We had flown together several times, and Janelle was always smiling, never in a bad mood, and didn't complain about serving and doing her share of the work with the passengers.  As I was placing the pretzels in their assigned compartment of the beverage cart, I heard a buzzing noise coming from the baggage compartment just behind the galley.  I turned to see which area of the compartment the noise was coming from, and I began climbing across the enormous pile of bags which had just been placed there by the guys on the ramp.  Pushing aside bag after bag, I continued to hear the noise…a pulsating, buzzing sound.  Suddenly I had my hands on the bag from which the noise was originating:   a small, green backpack which was not only buzzing but had a slight vibration.  I started to pick it up and then remembered our last recurrent training about bombs.

"Jesus Christ, " I said out loud, as I simultaneously grabbed the telephone off the wall mount in the galley and waited for someone in the cockpit to answer.

"What's up, Margo?" I heard Jonathan ask politely.

I was so relieved to hear the Captain's voice instead of the First Officer, who I did not know.

"One of the bags in the baggage compartment is vibrating and making a buzzing sound…it's really weird.  It doesn't sound like an alarm clock or a game."

"I'll be right back," he said.  "Don't pick it up."

"No problem," I answered.  Just like every flight crew member after 9/11, I was still jumpy, still nervous at times, but I was not really afraid anymore, and I was quite stoic in my resolve to continue flying and not to quit as many of my co-workers had done.

Jonathan's friendly smile was reassuring as he walked to the galley and peered through the curtain which separated us from the baggage compartment.

I pointed to the backpack which was still vibrating and buzzing, looking quite sinister.

"Shit," he said. "I'll be right back."

He turned and went back to the cockpit; it was only a minute or two before he returned with a concerned look on his face.

"They are going to clear the area around the aircraft; then they want us to throw it out onto the tarmac."

"I didn't think I was supposed to pick it up?" I asked him with a question in my voice.

"Apparently this is just a precaution," he said.

"Okay. Do you want me to do it or do you want to?"

"Ladies first."

"Gee thanks," I laughed nervously. We waited until we had a thumbs up from the ramp guys, and I threw the backpack out the galley door.

Nothing happened.

Two security guards came and picked up the green backpack and took it away.

"Well, I'm glad that thing's gone whatever it is," I said.

"Hopefully it's just a kid's game or an alarm clock like you thought, but don't worry, you did the right thing, " Jonathan said reassuringly. "I may need you to help me fly the airplane now; Paul cannot take his eyes off Janelle."

"I noticed that," I giggled. "Uh-oh, time for you to get back to your cage; here come the passengers."

"See you in a bit," Jonathan said over his shoulder as he walked up the aisle to the cockpit, away from the "herd" as we often called the never ending stream of people we transported during the killer four-day trips aboard the ATR.

I saw Janelle come through the small hallway that separates the cockpit from the passenger cabin of the ATR; the hallway always gives me the creeps just like everything else about the ATR. Baggage is stacked on either side and if the light is not turned on the flight attendant has to walk through a pitch black hallway from the cabin to the cockpit during flight, and also on the ground once the baggage door is closed .

"Paul is really cute….and he's nice too," smiled Janelle.

"We've got two days left on this trip, and an overnight in Panama City, home of the world's most beautiful beaches," I teased her.

"How do you know they are the world's most beautiful beaches," she quipped, " you sound like a tourist commercial."

"Because I was born there, I have traveled all of my life, and for some reason, the most beautiful sugar-white sand and emerald and blue water exist in the Florida Panhandle, known by locals as the "Red Neck Riviera," I laughed.

"Well if Paul is as interested as he seems to be, I may show him some brown sugar instead of your sugar-white beaches," she laughed teasingly.

"Oh, my," I exclaimed, "that must have been some introduction in the cockpit.

"I have my ways," Janelle smiled and she made her way to the front of the aircraft to the Flight Attendant B seat just as the first passengers were coming up the stairs .

"Why in the world do we board this aircraft from the back of the plane," a petite brunette asked as she tried to catch her breath after carrying her bag up the stairs in the stifling, humid Atlanta heat.

I almost said, "because everything is ass-backwards on this horrible airplane," but instead I smiled sweetly and replied with some nonsense about the design of the aircraft. I really felt sorry for her because the wind was blowing her hair almost off her head, she was struggling to hold her skirt down, and all the while being sandblasted with the sand from the hot, black asphalt of the tarmac. I reached down and helped her lift the bag past the last two steps, and she smiled gratefully at me as she began to make her way to her seat.

I continued assisting the passengers aboard, helping them stow their bags, and I also pulled the forms down from the galley cabinet which I would need for writing the Unusual Occurrence Report for the flight attendant supervisors regarding the buzzing, vibrating backpack I had flung from the galley door. I was happy for this trip to be almost done, but I was not even excited about being in Panama City for the night as I usually would have been because Lee was not going to be at home and I was just going to stay at the hotel with the others. He was gone to a NASCAR race for the weekend with his best friend James who is the owner of a NASCAR racing operation in Spartanburg, South Carolina.

As the passengers took their seats, Janelle made her way through the cabin counting the number on board, and when she got to the galley she sat down in my jump seat for a moment.

"God, my feet hurt," she complained. "Why in the world did I wear these new heels for the first time on an ATR trip? I should have worn them a few hours at a time going to church or something until they were broken in. I wonder if anyone would notice if I just took them off and walked around in bare feet?"

"Probably," I said glumly, "and then you would get an occurrence from the flight supervisors if someone reported you."

"You're right," she agreed, "and with some of the attitudes I've already encountered from this group, you can bet they would fill out that damn little Delta passenger comment card in the Sky magazine."

I saw the ramp guy coming to the steps to get the paperwork, and I reminded Janelle that she would need to go to the cockpit to get the completed paperwork from the crew before we could leave.

"I need to find an excuse to show Paul my new dance move," she laughed as she turned to go back up the aisle. As she did, she moved her butt in the most suggestive way possible, pulling her cheeks together and then shaking them rapidly, lowering herself almost to the floor before she got into range of the first row of passenger seats. She then walked completely normally the rest of the way up the aisle and disappeared into the cockpit.

"Oh, my God! I can't imagine the muscle control it takes to do something like that," I laughed to myself. Janelle was just plain fun to fly with, and when you are stuck with someone you may or may not like for four days of hard work in the confines of a sixty-six passenger aircraft, it helps to have someone with a little personality who smiles occasionally. This definitely promised to be an interesting final two days of the trip.

I was laughing as the ramper came up the stairs for the paperwork and I told him it would be just a minute before the crew finished up their calculations. I handed him a bottle of water, and he smiled gratefully at me, leaned against the doorway and drank the water in one gulp.

"It's hot today," he said as he wiped the perspiration from his brow. I handed him a paper towel from the galley and he wiped his face, just as Janelle came back down the aisle with the final paperwork from the crew.

"Thanks, have a safe flight," he said as he bounded down the stairs, and I began closing the door, watching the heat, sand , and wind disappear as the door sealed shut.

Janelle was walking the aisle making sure all bags were stowed underneath the seats and simultaneously running her hands along the overhead compartments to make sure they were closed. I removed the telephone from the cradle and began the safety announcement once she had returned to the front and began holding up the demo items: the seat belt, oxygen mask, and life preserver.

After letting the Captain know we were "secure for takeoff," I took my seat and Janelle and I waved to each other as the taxi and takeoff began. Her seat faced the passengers and my seat was against the galley wall behind the last row of passengers, where thankfully, no one could see me unless they turned completely around to look at me.

As soon as we reached ten thousand feet and I made the announcement to passengers that portable electronic devices could be used and that our beverage service would soon begin, Janelle was out of her seat and walking to the back.

"I'm gonna take Paul a Mr. Pibb," she said. "That's what he likes."

"Well call and ask if Jonathan wants anything before you go up there; after all, he is the Captain," I giggled. "He might feel neglected if you only take something to Paul."

Janelle picked up the phone with an apologetic smile and asked the boys if they wanted anything.

"Captain wants a black coffee and some Biscotti cookies; Paul just wants a Mr. Pibb."

"Go ahead and take it up," I laughed. "I'll get the carts ready while you are gone." As she went to the cockpit, I began pulling the two carts out and lining them up to hook together for us to serve the passengers. We would push the cart all the way to the front and begin with Row One, working our way back to the galley as we finished for cleanup, and also to hide out and eat our own lunch out of the sight of the passengers.

Most of the passengers on the way to Panama City were tourists, and many wanted beer, wine, or cocktails, and so I always tried to make sure I had plenty of change as well as lots of ice on the beer and white wine. Happy passengers were the best ones to fly with, and I always did my best to make sure they were happy whenever it was humanly possible. Flight attendants do not have to deal with cash anymore on flights because airlines only take credit cards, but in 2003, we still could not take credit cards on ASA, and flight attendants were required to have 20.00 in cash as part of our required duty-in equipment.  If you didn't have it, you could receive a demerit if the supervisor found out .  Other "equipment" included a watch with a second hand, the "wings" pin with our name on it, a flashlight, the flight attendant manual, and a complete uniform, including a scarf and pantyhose for females, and a tie for the guys.

The flight to Panama City was a little over an hour on the ATR, and by the time we finished serving, cleaning up, and having a few moments to gulp down our own lunches, we were ready to begin initial approach into the Panama City Airport.  This would be a twelve hour overnight in Panama City, and after only having to fly one leg today, I was ready to settle in my room with my book for a while, and then maybe walk across the street to the Panama City Mall and do a little shopping.  I was disappointed that Lee would not be home tonight, but I had also had a really hard two days with five legs on the first and four on the second, and my feet and legs hurt from standing so much. ( Crews stay on the beach in Panama City for overnights now because there is a new airport which is actually much closer to the beach than the old airport where we landed then. )  The Holiday Inn Select was the crew hotel then, and although it was dated, it was clean and had an indoor swimming pool and hot tub, a nice bar and restaurant, and on the floor where the crews were housed, the bathrooms had red bathtubs and red wall mounted telephones.  Flight crews found this a source of humor, and anytime you heard someone talking about the "red telephone" you knew they had been in one of the famous bathtubs in Panama City.

As we were walking through the terminal of the Panama City Airport, I ran straight into five members of the Phoenix Racing NASCAR race team owned by my husband's best friend, James

Finch  They waved me down, and I walked over to say hello.

"Where are you guys headed?" I asked as I hugged Johnny Allen, one of the team members.

"We're on our way to Atlanta and then to Las Vegas for the race," Timmy smiled. "Where are you going?"

"I'm just here for the night. Lee's already gone with James to Las Vegas, and so I am staying at the hotel with the crew."

Just then I noticed that both Johnny and Timmy had little stick on name tags on their shirts that said, "My name is Kevin Davis."

"What in the world?" I demanded pointing to the name tags.

"Lee told us about the guy that all of the flight attendants are looking for. We figured it was worth a try," Johnny laughed as they hurried to the gate.

"Oh, my God, " I laughed as I followed the others to the shuttle waiting to take the crew to the hotel.

As we started up the elevator to our rooms after checking in, the guys were talking about getting a cab out to the beach for dinner, but I told them I had seen the beach before. Janelle said she was "in," if they would give her a few minutes to change. She winked at me as she went in her room.

I didn't see them until the next morning as I was headed down the hall, still in my pajamas, to the vending machine for a Diet Coke. As I was going back into my room, Paul *and* Janelle came out of Paul's room together, in their flight uniforms, headed downstairs for breakfast. I looked straight ahead and mumbled " morning," as I inserted my room key into the door and dove into my room, laughing hysterically as soon as the door closed behind me, thinking of Janelle's comment about "brown sugar." For once I was eager to board the ATR and get started on the trip back to Atlanta, just to hear all of the details from Janelle, and I was certain she would tell!

## 13  Key West Wedding

"Six more days, " I thought as a pulled my black bag down the C Concourse, headed for my gate for the last trip before I would be getting married; it was an ATR trip, and I was hoping that I would be flight attendant "A" so that I could be in the back of the aircraft instead of the front.  I still hated sitting in the front seat which folds down facing the front row of passengers.

With barely three months since 9/11 the numbers of passengers walking through the airport still seemed small compared to the normal crowds which had usually poured through the Atlanta concourses rushing from gate to gate, grabbing a meal from the fast food vendors and racing for the trains to whisk them from one concourse to another.

I had started flying again on September 16th, and many of my colleagues were still at home, too afraid or too depressed to come to work; my boyfriend Lee had proposed to me on the morning of 9/11 as I got off my flight, and I was so excited planning my wedding that terrorists were not even part of my thoughts today.

My main worry was getting home to Panama City after this four-day trip and having time to pack and get ready to go to Key West for my wedding on December 29$^{th}$. I had arranged my vacation, dropped and traded some trips, and I would be off until January 15$^{th}$-- I would have two weeks off after the wedding, and a few days before.

As I walked along the concourse I thought about my original plan of having a dinner at the Ernest Hemingway House for our reception after the wedding at the old stone church, a Methodist Church on the corner of Eaton and Simonton Streets in Old Town, Key West. Lee's best friend, James Finch, who owns a NASCAR racing team, a jet, *and* a yacht in Key West, had called and offered his 114-ft yacht, the *Phoenix* for our reception, and so I happily agreed, and about seventy-five of our friends were flying to Key West for the wedding. Lee's two daughters would be bridesmaids, my daughter would be the maid of honor, and they were also bringing friends, all of who would be staying in the bed and breakfast inns surrounding the church .

"*I cannot believe this is finally happening,*" I thought as I showed my ID to the gate agent and started down the stairs to the aircraft. "*There it is….the goofiest looking airplane in the fleet*, I thought , as I walked across the tarmac, and a goofy-looking First Officer was walking around doing his "walk around" inspection. I had never seen him before, but honestly, where do they find some of these pilots. He looked at the tires, then he kicked one of them, and at the same time, touched the underside of the plane as though he were stroking the belly of a large bird. Then he spit on the ground.

"*Why do men always feel they have to spit or scratch their crotch when they are doing something mechanical?*" I wondered. Just then he scratched his crotch and hiked his pants up in the back. I giggled out loud and yanked my bag up the stairs as I climbed to the galley. The food service employee had put a bag of ice in the drawer and was leaving, and the cleaner was walking up the aisle with her air freshener making small puffs of orange smell as she moved toward the galley. I was sure she had not wiped down the trays or reached into the seat pockets for the nasty trash passengers stash in there –the cleaners never do. Then when the passengers fold down their trays when we serve snacks and drinks they look at me like it's my fault there are crumbs and drink circles left from the last inconsiderate passengers who sat in their seats.

*"How will I ever get everything packed and ready for the wedding in just four days?"* I silently asked myself as I began walking through the cabin arranging the seatbelts, making sure all of the safety equipment and first aid kit were in place, which by now included a defibrillator , the latest piece of equipment I had received training to use.

As I was completing the safety check and starting to put ice in a cup for a Diet Coke for myself, the other flight attendant came up the stairs, and she smiled at me as she dragged her bag up the steps. I couldn't imagine what it must be like not to have to climb stairs to an aircraft like the Delta flight attendants who walked onto their flights with hair and makeup intact instead of being blown to Hell and back by sand while the Atlanta heat melted the makeup off your face.

"My name is Katie, " the petite brunette said, "what's yours?"

"Margo," I said as I smiled back at her. "I'm looking forward to getting this flight over with and getting home for Christmas and vacation, how about you?"

"I'm afraid that's not going to happen, " she said. "I am on reserve, and I have a feeling I will be flying on Christmas Day."

"Sorry, " I said. "I flew last Christmas, but it wasn't too bad—my fiancé flew to Toronto on the flight with me and we were able to spend Christmas together. We even got snowed in and got to spend a few extra days together!"

"That was lucky," she giggled. "My boyfriend is still mad at me for taking this job. He didn't realize I would be gone so much, but I have always wanted to be a flight attendant, and I couldn't believe it when they called and offered me a spot in the class. I took it without even talking to him about it."

"It can be difficult," I said, remembering all of the arguments with Lee and days without calling during my training as a flight attendant, and especially during reserve when I never got to go home.

"Are you commuting or living here in Atlanta?" I asked.

"I'm commuting, and from one of the hardest cities to get in and out of so I'm told," she said. "Myrtle Beach."

"Oh, God, that *is* tough. I live in Panama City which is almost just as bad," I commiserated. "I'm getting married four days after this trip, and I'm just praying that the trip goes smoothly, we don't get delayed or extended, and I can get a flight home in time to get to Key West."

"Wow, congratulations," Katie beamed sincerely. "Maybe everything will just go perfectly and you can get to the church on time ," she laughed.

I was laughing with her and about to offer to pour her a Coke when I saw two flight attendants heading for the bottom of the stairs of the ATR. They were smiling and laughing as they walked toward us, and suddenly I recognized them and waved. Sheila and Kathy were veteran flight attendants who always flew together and always flew on the ATR; Kathy was divorced and Sheila was married to a Delta pilot but had a boyfriend who was a pilot on the ATR. She always bid trips when she knew she could fly with him, and Kathy always went with her .

"What are you two up to? " I laughed as they approached.

"We are coming to kick you off your flight and get you on the way to your wedding," Sheila said.

"What!" I exclaimed. "You are kidding me, right?"

"Nope, " said Kathy. "Sheila wants to fly with her favorite Captain and I don't want to be alone on Christmas. We went to the flight supervisor and asked for this flight, reminded her about your wedding and vacation, and you my dear, are now relieved of duty!"

I grabbed both of them in one hug and began to jump up and down screaming excitedly.

"What do I do?" asked Katie. "Should I call crew scheduling?"

"Are you kidding me, " said Kathy. "Just grab you bag and run like Hell before they know you are not flying." "Merry Christmas!"

Katie might have been new, but she was quick. No one had to tell her twice. She started down the steps with her bag before I could even get might out of the overhead bin where I had stowed it.

"I don't know how to say thank you," I said as I brushed the tears out of my eyes and hugged them again.

"Oh don't go getting all emotional. You are just getting married," Kathy laughed. "Now get out of here and go to Key West. "

As I walked to the Concourse I passed Captain "what's his name" and he smiled at me and waved. I smiled back; I never have understood why married people cheat on each other, but I was too happy to worry about anyone else or be judgmental today. I headed down the "C" Concourse to check the flight board. I had not counted on leaving for Panama City this morning, and I wasn't sure when the next flight was leaving, or if there might possibly be a seat. I didn't care; I was delirious with happiness. "I will rent a car and drive if the flights are full," I thought to myself.

My eyes quickly scanned the lighted Departures board in the center of the Concourse and there I saw it, PFN, Gate 36, "ON TIME" 10:30 AM. I glanced at my watch which said 9:50. I raced down Concourse C for Gate 36, all the way at the end where there were usually two or three flights boarding at one time. I ran to the counter, laid my ID's on the counter, and begged, "Could you please list me as standby on this flight. I just found out I could go home, and I am getting married."

"Congratulations," smiled the pretty African-American gate agent. "Let's see how it looks."

"Thank God, " I thought to myself, "*a nice gate agent; miracles just never cease on this perfect day.*" I suppose any job has unpleasant people, but it sometimes seemed that every bitchy woman in Atlanta applied for the gate agent positions with ASA.

"Ok, you are listed," she said, "and I think you will probably make it. We will be boarding in about ten minutes, and I will let you know as soon as I can assign stand- bys."

"Oh, thank you so much!" I said, and I felt tears in my eyes again. I didn't know what was wrong with me; I couldn't seem to stop crying today. Since 9/11 tears and emotional outbreaks were not uncommon for me and other flight attendants, and everyone seemed a little nicer, a little more concerned for each other than before. I sat down in a seat as close to the gate as I could manage and waited for the flight to begin boarding. I wanted to call Lee, but I decided to wait until I was actually on the flight before I surprised him. Looking around at the passengers waiting for the flight, I began to mentally count them to see if there were 66 people , the number of seats on the ATR. Of course some of them might be waiting for the TRICITIES or COLUMBUS flights which were also getting ready to board soon, I reminded myself.

"Ladies and gentlemen, we are ready to begin boarding the flight for Panama City, Florida. Please have your boarding passes ready, and if you have a roller-board carryon you will need to get a pink valet tag as you board the aircraft. Your bag will be returned to you at planeside when you arrive in Panama City," the gate agent said as she began the boarding process.

I watched anxiously as the passengers began to move toward the boarding door. The gate agent was efficient and pleasant to each passenger as they moved through the line, and I noticed she was wearing one of our Delta Christmas scarves. During the holiday season the company offered a new Christmas design each year; the pilots and male flight attendants could purchase ties and the flight attendants scarves. Many of us had been wearing American flag scarves and patriotic pins since 9/11, and it was nice to see a reminder of something normal, even if it was a scarf with whimsical holiday teddy bears on it.

As the passengers were moving through the line, I noticed one slightly bald, heavyset man in a Christmas t-shirt as he was finishing off a piece of chicken from a Popeye's Chicken box. He meticulously gnawed on the chicken leg, inspecting it after each bite to see where to take the next bite.

*"Honestly,"* I thought to myself, *"how can someone sit in the middle of the airport eating a greasy chicken leg without feeling the least bit self-conscious."*

The man now placed the end of the leg in his mouth like a lollipop and actually sucked the top of the bone before dropping it into the box, wiping his mouth with the back of his hand, and then wiping his hands on the back of his pants after he dropped the box onto the table next to his seat as he moved to get into line . The line was beginning to come to an end, and as I waited in anticipation for the call for stand-bys, suddenly four people came running through the concourse, waving their boarding passes, as one shouted breathlessly, "Are we too late for Panama City?"

"Oh no," I thought. I had been counting and I knew we were close to a full flight. People were starting to fly again now that the holidays were here, and even though most flights were still not nearly at capacity since 9/11, those headed to warm, beach destinations like Panama City were definitely starting to fill with passengers, who while nervous and fearful, were ready to start traveling again. I watched miserably as the four passengers boarded the flight, and then suddenly heard the gate agent say, "all stand-bys have now been cleared for boarding."

I leaped out of my seat, grabbed my bag, and traded my standby NRSA ticket to the pretty gate agent for a boarding pass with a seat assignment.

"Have a great wedding, sweetie," she said as I smiled gratefully at her. I wanted to grab her and hug her, but I said, "thank you," and continued down the stairs to the tarmac and boarded the ATR, seat 15C, almost in the back, allowing me to be almost the first to get off the plane when we arrived in Panama City.

As I settled into the seat after stowing my bag in the overhead bin, I finally took out my phone and dialed Lee's number. "Hello, this is Lee, please leave me a message," said Lee's voicemail.

"Hello, Lee. This is Margo, my trip was cancelled, and I am on my way home. Love you!"

I closed my eyes, listened to the A Flight attendant begin the long announcement, the noisy sound of the prop, and before we ever took off I was asleep, dreaming of Key West.

Margo Deal Anderson

## 14  Hawaii in 72 Hours

Hurricane Ivan was headed straight for Panama City , and as the weather forecast became more inevitable, the long-awaited announcement came from Bay District Schools that they were closing for the next five days.  By now, I was flying part-time for ASA , mostly on weekends and as little as I could possibly get away with, and teaching school again.  My husband Lee and I were enjoying flight benefits, and I flew to Dallas twice a month to fly the required hours for a part-time flight attendant.  I had changed my base from Atlanta to Dallas because there were fewer flight attendants and my seniority was higher, allowing me to be on the part-time roster, even Part time A, which meant I got first choice of the trips on the schedule which were split in half.

I had ridden out the last hurricane which directly hit Panama City, Hurricane Opal, in 1995, and I was not eager to experience the flooding, lack of power, and deadly winds of this one, which was predicted to be just as bad. I called my husband, Lee, to let him know I was on my way home from school. We lived in a small lake cottage about 20 miles north of Panama City instead of on the beach as we had when we first married, and our location was perfect for the "spin-off" tornadoes which always accompany hurricanes. As soon as I walked through the door, I noticed that Lee's bag was packed , sitting by the door. I laughed, and asked, "Where are we going?"

"Cincinnati ," he answered. He had already been on the computer looking for seats on an outbound flight.

"Great," I answered, and flew up the stairs , grabbing my flight bag out of the closet as I went. Lamb Chop, my two-pound, white teacup poodle began to whimper as soon as he saw my flight bag. Next came his usual leap off the floor to the bed where he promptly got into the middle of the bag looking at me defensively, with his best, "where are you going now and why are you leaving me again look."

"What are we going to do with Lamb Chop?" I yelled down the stairs. " I called the Beach Pet Motel already and they are full; the other one in town is not accepting any pets."

"We can take him to the one in Chipley," Lee said, "you know, the veterinarian with the farm." "He will be further away from the storm and safer there." Chipley is about twenty miles north of our lake house, a small agricultural town about fifty miles south of the Alabama border.

"I don't want to leave him," I said, knowing that the argument was futile; I knew he would be better in a safe dry kennel than flying with us, maybe hanging around in Atlanta for hours waiting for the connecting flight.

"He's not going, Margo."

"I know." I looked sadly at Lamb Chop and began to pack his little bag to go to the kennel. " I will be back before you know it," I told him sadly. His face told me that he knew I was lying and that he had heard this story before.

"Why Cincinnati for five days, of all the God-awful places to pick?"

"Hey, we have to go someplace; the flights are full. If we can get there, you can pick another city tonight."

He was right; the flight card would basically take you anyplace in the world if there was a seat, and often sitting in First Class. The only drawback was having enough money to stay in a hotel and eat in restaurants when you got to your destination; this puts a real damper on traveling for most beginning flight attendants.

I finished packing, we loaded the car and began driving toward Chipley to drop off Lamb Chop, and then we would continue to Dothan, Alabama, to catch our flight from there. I called my Mom and Dad to encourage them to evacuate with us, and they were determined not to leave. The last time evacuation was called for, people were not allowed to return to their neighborhoods and homes for over ten days , and one of the best things about having stayed, was that we were at least in our homes and able to start cleaning up and getting back to normal as much as possible using generators. I told them that we were leaving, but they still would not go. My parents both had flight privileges as well, but they felt safer in a hurricane than on an airplane, especially after 9/11. I knew arguing would not help, and I asked them to be careful, take care of themselves, and I would stay in touch.

At this point, I was not feeling good about the trip at all; I was abandoning my dog, and now my parents, but I also knew that our location was not safe, basically in what is known in the south as "tornado alley" in a mostly glass chalet on stilts beside a lake. My parents were in a flood zone, but in a solid house which had withstood much worse, and because my father had been a city commissioner in the past, they at least had people nearby who would be accessible to them if they had to evacuate, since they lived only a few blocks from city hall.

We finally arrived at the Dothan airport, and we began our flight to Atlanta and then to Cincinnati; we had reservations at the airport Hilton, and upon checking in found that we were paying almost $300.00 per night for a view of the Cincinnati airport.

I immediately began checking for seats to another destination.

"How about Hawaii?" I asked .

"Are you kidding me?" Lee said.

"No, really, there are four First Class seats open and its leaving in six hours."

"Hell yes," he grinned. "Let's go."

"Oh my God, " I shouted. "I've always wanted to go to Hawaii!"

"I haven't been since Vietnam, but I can tell you, it's one of the most beautiful places you've ever seen," Lee said.

I listed us both for the flight, then called the front desk to see if we needed to register for the shuttle from the hotel to the airport, which would now leave us only a few hours of sleep.

"I have nothing to wear in Hawaii," I lamented as I began to look through my bag.

"Don't worry about it," Lee said. "We are only going to be there for a couple of days. We can buy something when we get there."

"Where do you want to stay?" I asked absently as I was perusing the travel guide which I pulled from my flight bag. Airline personnel not only have standby flight privileges, but are also given great discounts at various hotels around the world with proper ID, sometimes as much as 50%. When we went on our honeymoon, we stayed at the El Presidente resort in Cabo San Lucas for half-price, and it was a beautiful, all-inclusive resort on the Pacific Ocean. I began calling out the names of hotels on Waikiki Beach, and suddenly Lee shouted, "That one."

"Which one?" I laughed.

"The Royal Hawaiian."

As I began to look closely at the amenities and photographs, I realized this was one of the oldest, most beautiful luxury hotels in Hawaii, located on the big island of Oahu.

"This is going to be really expensive," I said as I continued to scroll through the photos, "even with the airline discount."

"I don't care," said Lee. " I walked through it when I was there during Vietnam; I was nineteen, and I couldn't even afford a cup of coffee. I promised myself that someday I would go back."

"Awesome," I said. "Give me the credit card and I will make the reservation."

A few hours later we were seated in First Class on Delta Airlines, headed for Hawaii. I was once again envious of the Delta Flight Attendants in their cute little grey culottes with turquoise sweaters. The announcement was a pre-recorded video instead of the loud, unbelievably long memorized speech we were required to do on the ASA flights. The Delta flight attendants did not like their new uniforms which had come out in 2000 just after I was hired; they wanted their navy blue ones back.

"At least their uniforms fit," I said to Lee. At ASA we had to wear navy blue pleated pants because the company managers thought the pleats were more flattering to some of their overweight flight attendants. The other pieces of our uniform included blue and white striped shirts (which made us look like we worked at Wendy's), a buttoned blue vest and a scarf. We also had blue skirts and blue dresses, but they were also cheaply constructed and ill-fitting.

The flight attendant knew we were non-revenue passengers from the manifest, and she was super-friendly, asking me where I was based. I confessed to her that I was an ASA flight attendant, but that I was based in Atlanta and was thinking of applying to Delta because I hated flying with ASA so much.

"You would be great with us!" she said. "Just do it…what are you waiting on?"

"I just hate to lose the seniority I have built up with ASA, and besides, my flying benefits are the same, and I really don't think I could bear to be on reserve again."

"Our reserve is different ," she said. "I have heard that ASA's is brutal."

She made mimosas for us, and offered breakfast choices; we talked for a few more minutes about the advantages of Delta vs. ASA, and then she moved on to the next row of passengers seated in Delta's amazing First Class section.

"Why don't you think about it?" asked Lee. "They certainly seem to have things better than that chicken-shit airline where you are."

"I will," I said. "The only thing is , that if I am accepted there is a nine-week training period and I might get based in Newark, which would just mean more time apart for us."

"Whatever you want to do is fine with me," said Lee.

Our seventy-two hours trip to Hawaii became a blur very quickly. The Royal Hawaiian was everything it had promised to be in the flight magazine and on the website. The room was beautiful and filled with pink towels, pink bathrobes, pink sheets, and pink spa amenities…Lee said it looked like a pink fairy had puked in the room. Romantic. I was ecstatic…I loved pink, I loved Lee, and I was absolutely in love with Waikiki Beach. Everyone always says it is overrated, but we had a wonderful time.

I had never realized that only the men sing in Hawaii; the women just do the beautiful traditional dances while the men sing and play the ukuleles . We visited Diamondhead, the somber, but beautiful Pearl Harbor Memorial, the pineapple plantations, and even went to one of the final shows of Don Ho, who passed away not too long after our visit.

I remembered Don Ho from his role in the original *Hawaii Five-0* series in 1969 when I was in the ninth grade. I thought he was so handsome, and at age 14 had a huge crush on him. What I didn't know was that Lee had actually met him during his time in Hawaii when he was serving aboard the Princeton in the Vietnam Conflict. Not only did he meet him, but apparently they pulled several good drunks together, and he said when his ship was in and out of Hawaii that he even visited his house. I wanted to meet him after the show, and we lined up with all of the others for autographs after listening to him sing all of his famous tunes, including, "Tiny Bubbles." As Lee and I approached, his face suddenly lit up as he saw my husband and he shouted, "Where you been?"

I turned in amazement to see my husband's equally delighted expression, and we were ushered quickly to the front of the line, where they embraced in a bear hug and began to talk excitedly. He really *knew my husband!* We took several photographs with him, and he autographed them. Later as we were talking, I learned from Lee that Don Ho had served in the U.S. Air Force accumulating thousands of flight hours throughout the Pacific basin, supporting the U.S. Air Force mission there. After speaking with this beloved entertainer and enjoying watching the reunion between he and my husband, I did more reading about him, and I hope his legacy as a dedicated member of our nation's military will be remembered always, alongside his life as Hawaii's prolific entertainer and beloved son.

After seventy-two hours of absolute heaven; we arrived at the airport for our return flight, and as I looked at the computer found that the First Class section in the flight we had booked was completely filled.

"Oh, no," I said. " We don't want to sit in the back of the plane for all of those hours." I quickly looked at the next flight and found that it had six available seats in First Class. I re-listed us as the flight we were scheduled to be on was boarding. We went to the bar to drink and hide out while they called the names of stand-bys who had been cleared. We listened to our names being called several times while drinking our last Hawaiian pina coladas, and then as our plane departed the airport, Lee looked at me and said, "Are you sure we will have seats on the next one?"  There was a slight panic in his voice.

"Of course, silly," I giggled.

"How long until the next flight?"

"We only have about two hours to wait, and that will be well worth waiting for.  I don't want to fly all the way to Los Angeles with my knees under my chin in a middle seat."

We ordered more pina coladas,  then walked through the gift shop and selected several souvenirs for my Mom and Dad, and then went to our gate.   Another forty-five minutes later we were sitting in seats 2A and 2B sipping champagne and getting ready to return home, chatting happily with the First Class flight attendant.

"So how did you like your first trip to Hawaii?" Lee smiled at me.

"Aloha, baby," I said as I snuggled against his arm.

Margo Deal Anderson

## 15  Chattanooga

"We apologize for your delay here in Atlanta. The estimated departure time for Chattanooga this evening will now be 8:45 p.m. due to maintenance."

"8:45!" shouted the man standing at the desk staring into the face of the gate agent.

"The scheduled departure for this flight was 4;30; we have changed gates, we have waited on a flight crew, and now the damn thing won't fly?  What's wrong with it?" demanded the man.

I sat with my crew bag in a seat near the gate, listening to the gate agent , and I smiled to myself as I wondered to what degree of hate for ASA the feelings of Chattanooga fliers had risen during the time I had been a flight attendant.  ASA had been featured in numerous front page news stories in Chattanooga, blasted for the level of service provided, including late flights, and cancelled flights to the eastern Tennessee city. I often wondered why the business travelers even bothered with the short flight; it was almost an even bet that they could have driven the distance round trip twice during the time they spent sitting in ASA's Concourse C of the Hartsfield airport on any given day.

After a grueling three-day trip on the ATR, I was waiting for my flight to Panama City, which would leave from this same gate in two hours. I was tired of the level of noise in the flight attendant lounge, and so I had changed clothes, grabbed a hamburger at Wendy's, and settled in with my James Patterson novel, trying to block out the conversations around me.

"Sir, there is no need to shout," the tired, slightly plump, female gate agent answered. She had shoulder-length brown hair, pulled back in a messy knot at the base of her neck, no scarf, and her blue and white striped shirt hung outside her wrinkled navy-blue trousers.

"Don't tell me not to shout; I have been sitting at this damn gate since three o'clock this afternoon, and now you are going to reprimand me for being angry?" he shouted again.

"Sir, if you will please be seated; I do apologize for your delay. Hopefully we will be able to get you on your way soon," she replied curtly before looking past him at the woman in line behind him. He moved to the side, but continued to stand at the counter glaring at her.

"He probably won't be flying at all today," I thought to myself. "He's going to find his ass kicked completely out of the airport if he keeps this up." I looked back down at my book and read the same page again. I couldn't concentrate on the page because I knew what was happening behind the scenes with this flight; I had seen it done before. ASA would send the aircraft which was supposed to go to Chattanooga to another city on a turnaround if they had a plane in maintenance or didn't have enough crews available. Then when the plane returned they would finally send the passengers on their fifteen minute flight to Chattanooga, mad, frustrated, and ready to fight. I knew things were not going to get better at this gate for at least two hours. Chattanooga was an easy flight for ASA to use as an "adjustment" flight to schedules because it was such a short run from Atlanta to Chattanooga; the ATR's were usually the aircraft for this flight, but as more jets were added to ASA's fleet, the flight became even shorter. If a flight was scheduled, say to Myrtle Beach, for example, and there was a maintenance or crew issue, it was a simple matter for scheduling to divert the Chattanooga aircraft to Myrtle Beach, and once it returned use it for the Chattanooga flight a mere two hours later. This type of abuse happened so often that ASA was featured in a front page story of the *Chattanooga Times*, blasted by the reporter for its poor record of on-time service, or for that matter, any kind of service.

As the time for my flight to Panama City drew closer, the Chattanooga crowd became more vocal, and I was honestly worried that ASA might divert *my* flight to Panama City instead of theirs. Then, just like clockwork, I heard the gate agent make her announcement.

"Ladies and gentlemen here in the gate area waiting for Delta Connection service to Chattanooga, Tennessee. We do apologize for the delay here in Atlanta, but we would like to announce a gate change at this time as we are experiencing maintenance problems with this flight. Your flight will now depart from Gate C 26. We ask that you remain in or near the gate area for further announcements."

I breathed a sigh of relief as I watched her begin to change the sign from Chattanooga to Panama City at our gate, but just then the man moved back to the center of the gate area and began yelling.

"What do you mean the damn gate has been changed? This is our second gate change in two hours' time, and I am not going to stand for it. I am a Platinum Medallion member of the Delta family, and this little two-bit airline with your piss-poor service is not going to delay me one more time. I demand to see a red coat right now!" he bellowed.

"Sir, if you will just calm..."

"I am not going to calm down. A red coat...NOW!!!"

I stared more intently at my Patterson novel and pretended not to be listening; like everyone else in the gate area I was hoping the guy would be successful in his demands, but I was doubtful. I had a feeling he was about to be leaving the gate area with an escort.

The gate agent looked past the man as though he were not there, and asked the woman behind him, "May I help you?"

He stepped between the counter and the next woman and screamed:

**"A RED COAT....NOW, NOW, NOW!!!"**

The woman backed away, the gate agent disappeared through the door behind her, and everything became strangely quiet as the man stood there with his face turning redder by the moment.

"Oh my God, I hope he's not going to have a heart attack or something," I thought.

Seconds later two security guards entered through a door from across the concourse and started directly for the man. With one on either side of him he began to walk down the concourse away from us, still loudly protesting about being a "Platinum Medallion……".

"Those passengers in the gate area whose destination is Panama City, Florida, your aircraft has arrived and once those passengers have deplaned and the crew notifies us that they are ready, we will begin the boarding process."

"*And that is that,*" I thought as I began reading again.
"*Chattanooga passengers screwed once again. SCORE: ASA: 1, Chattanooga : 0* "

Margo Deal Anderson

## 16  Freedom Fries and French Wine

My husband and I were traveling in Germany, he as part of a business trip with his company, Berg Steel Pipe, Inc., and I decided to travel with him using my Delta flight pass card.  He and his colleagues boarded the aircraft when their seating zones were called in Atlanta, and I waved at him when he passed through the gateway leading to the plane.  I was positive I would have no trouble in securing a seat; I had just checked the computer and there were 16 available seats in the coach section and 4 available in first class.  I was secretly hoping for the First Class seat, and I had made sure not to wear jeans , attire which would automatically exclude me from sitting up front due to a company policy that required employees to follow a dress code when flying non-revenue and wishing to sit in the First Class cabin.

Lee, my husband, had a worried expression on his face as he waved at me; he never relaxed when I used my pass, and he had offered to purchase a ticket for me, but I am always a bit of a risk-taker. Typical of the ATL gate agents in the days before the non-revenue seats were posted on a computer screen for everyone to see, this particular gate agent smugly reminded non-revenue passengers that she would notify us when seats were cleared and not to bother her again (one pilot had gone to the counter three times to inquire about seats, and she was noticeably irritated).

I had counted the First Class seats, and I knew there were still four seats available. I was becoming a little angry at this point, because I wanted to have a few moments to enjoy the "perks" of sitting up front—a nice cocktail, placing my luggage in an uncrowded overhead compartment, and getting comfortably seated several minutes before takeoff. Now, thanks to Miss Congeniality at the desk, I would probably have to run down the gangway , toss my bag into the compartment, and hear the Delta flight attendant tell me regretfully that there was not enough time for me to have a cocktail before we began the safety announcement .

At exactly ten minutes before the flight closed, the gate agent cleared all standbys, and I was given seat 2B, a First Class seat. "*Yes, Virginia, there is a Santa Claus,*" I thought as I handed the boarding pass to the agent scanning the passes. I waved at my husband as I boarded, and then I quickly turned left and entered the cabin known as "the front", stowed my carryon bag, and slipped into 2B. As I had predicted, the flight attendant sadly told me that she wouldn't have time to serve me until we were in the air. I decided not to be angry, that I could make up for not having an on the ground cocktail by having more in the air once underway.

After we reached cruising altitude I walked back to the main cabin to tease my husband who was traveling with his company on a corporate ticket, which, of course, was not First Class. He waved me away, saying, "Don't talk to me."

"Not even if I am bringing you a Crown and water with lots of ice?"

"Well, ok," he conceded.

I sat on the arm of his chair and said, "I'm sorry. I will trade seats with you just to show how much I love you."

"No that's ok. I should have listened to you and flown on my standby pass, but I was just afraid I wouldn't make the flight; I have a really important meeting that I have to attend, and after that we will just be able to have a good time."

At that time, Lee worked for Berg Steel Pipe, Inc., a division of EuroPipe, and although I had been with him on business trips to Germany before, this was the first time we had been since 9/11, and many passengers were still nervous about what had happened and also about what the United States response would be. At this point, there had been no retaliation or attacks of any kind by our government.

I talked for a while with Lee and with one of the purchasing assistants and her boyfriend who were traveling with us, Rena and Donald. They had only been dating for a short time, and Lee had actually introduced them. He and Donald had been friends for a long time, and we were both surprised to see that Rena had invited Donald on this week long trip since they had not known each other very long.

*"Must be going pretty well,"* I thought to myself as I returned to my seat to settle in with the inflight movie which was just beginning.

Nine hours later we cleared customs in Dusseldorf, and we rented a car to begin our eight day trip in Germany. I am not particularly fond of all of the German sausages, but I do really like their potato soup and also the Muscovy duck which is served rare over various fresh salads.

Lee's first meeting was in the town of Dunkerque, a beautiful coastal city on the northern shores of France, which still bears the scars of the bombing raids of WWII. While he and his assistant, Rena, were attending meetings, Rena's boyfriend, Donald and I, were left to explore the town of Dunkerque on a Tuesday , after checking out of our hotel. The plan was to meet Lee, Rena and the German entourage at the hotel bar at approximately 5 p.m. that afternoon.   What we had no way of knowing, was that the entire town was closed on Tuesdays, except for the bars. Donald and I knew each other, and while we were not good friends, we were certainly more than acquaintances, and so we began walking the streets of Dunkerque looking for something to do. We found a few shops open, and we bought some souvenirs and also some gifts for Lee and for Rena. Unfortunately, most of the stores in Dunkerque were closed on Tuesday for some reason, and so Donald and I , to avoid freezing, began to bar hop. Our luggage was stored in the lobby of the hotel where we had all checked out, and we would meet the others to return to Germany later that afternoon.

At 5 p.m. my husband, along with Rena, burst through the door of the hotel bar to find me standing on a table leading the patrons of the bar in the French National Anthem, La Marseillaise, having a grand time.

"Vive, Lafayette!" I shouted.

"Vive, Margo!" The bar patrons were crowded around me and the guy playing the piano, as Donald leaned against a chair to stand up as he lifted his beer in the air shouting, "Lafayette, Lafayette!"

I began to direct the patrons again, using exaggerated arm motions and singing at the top of my voice,

*"Allons enfants de la Patrie, Le jour de gloire est arrive!*

*Contre nous de la tyrannie,L'entendard sanglant est leve',*

*Entendez-vous dans les campagnes...."*

I stopped singing suddenly as I looked directly into Lee's angry face standing in the doorway with Rena.

Rena apparently thought the worst of her boyfriend and I, and even Lee seemed suspicious as he stared at me. I climbed down off the table, amidst the protests of about thirty singing Frenchmen, and Rena punched Donald as she wheeled around and went out the door.

We boarded the van with all of the German business associates in complete silence. I tried to explain about every store in the town being closed and nowhere for us to hang out to avoid freezing in the 20 degree temperature except in bars, but to no avail. I began to get more and more angry, as I thought about Donald and I, left on the street after we all checked out of the hotel. Everyone had assumed that Donald and I would spend the day shopping and sightseeing. Tears welled up in my eyes as Lee continued the silent treatment. To make matters worse, Donald ran, stumbling, to the front of the van and the driver had to stop while he threw up. He never looked at me when he crawled miserably back onto the van and sat down in the seat next to Rena.

No one spoke for the next two hours as we drove from the northern French coast back to Germany

As we rode in silence, it became apparent to me that Donald and I were being accused of some sort of impropriety, when in truth, we had just been hanging out in bars because the others had not bothered or had the foresight to arrange an extra day at the hotel where we might have been more comfortable. I began to get angry. We stopped at the equivalent of a rest area/gas station after driving for so long in silence, and as I started for the restroom, Rena brushed past me in a huff. I waited for her outside the stall of the ladies room and stood in the doorway of the only exit, and then got right in her face:

"Let me tell you something you little bitch; I have watched you get drunk for the last three nights and fall all over the corporate managers making an idiot of yourself, with no worry of how you are representing Berg Pipe; how dare you insinuate that I would be interested in your little pantywaist boyfriend from Muscle Shoals, Alabama. You are trying to cause a problem between my husband and I, and I can show you the meaning of redneck, because I come from a long line of them. You either start speaking to me and acting like we are at least cordial acquaintances, or I will kick your little ass all over this German truck stop. " With that, I turned to walk back to the van, and I ran straight into Lee, who had heard everything and was laughing out loud.

She never said a word, but rather, marched right past us and went and got back on the shuttle van.

I sat down next to my husband as we boarded the van with eight pairs of eyes watching us, and I spoke to him very quietly.

"Look, there is more to this story than meets the eye. So you either start speaking to me again and realize that these two people are crazy, or I will buy myself a ticket home. Screw that little Alabama bitch; I didn't do anything wrong unless staying inside the bars all day to keep from freezing to death is wrong. I will admit I had too much to drink, but it's not my fault you left me on the streets with nowhere to go. How was I supposed to know the entire town closes on Tuesday?"

Lee moved closer to me in the seat and took my hand. When we arrived back at our hotel in Germany, we went to the restaurant and Lee even called Donald and Rena's room to see if they wanted to go with us, but they didn't answer their phone. After a bottle of wine and dinner we were fine, and we went back to our room for a very romantic ending to a very tumultuous day.

At approximately 3 a.m. I awoke to a loud shrieking noise which sounded like a cat. The noise started at a low pitch, and then rose to a high soprano shrieking, almost like fingernails on a chalk board.

"What the hell," Lee began, and then, before he could say more, the shrieking started again.

"Oh my God, " I said, "it's Rena and Donald next door."

For the next hour we listened to the most god-awful, unearthly wailing and shrieking one could imagine as the two of them made love. Rena's voice would start at a low pitch, and then it rose, louder and louder, sounding exactly like a cat on a moonlight night.  Then the headboard of their bed began to bang against the wall in a rhythmic noise, coinciding with the rise and fall of the cat wails.  Lee and I laughed hysterically until we cried, and suddenly everything was alright between us.

Just before I fell asleep, Lee said, "Look, the whole thing was worth it just to see how pissed off the Germans with EuroPipe were at you for singing with all of the French guys in the bar.  I am tired of them anyway.  I thought you were going to start a riot— how in the Hell did you get them all to sing with you?"

" I don't know...I was just picking out the tune of La Marseillaise on the piano, when a guy sat down and started really playing it, and I started singing.  Then someone bought me a shot of bourbon, and for some reason I yelled out, "Vive, Lafayette," and it just kept getting louder and more fun," I giggled, as we cuddled together under the thick down comforter against the cold German winter night.

For the next two days Lee attended conference meetings and we went out to dinner at night; we toured various tourist destinations in Germany and France, and we had a wonderful time.  Finally, on the day before we were scheduled to leave we were having dinner with a large group of the Germans, and as we were having dinner, someone turned on a television in the restaurant, and the BBC  announced that the United States had begun bombing Iraq.  I suddenly felt very afraid, and I told Lee I wanted to go home.  He agreed.

The next morning we began making arrangements to leave, and as we arrived at the airport in Germany to board a small commuter airline which would take us to Paris, protestors were lining the road with anti-American signs.

"Great," I thought, as we got out of our cab and went inside with our bags.

We were both flying on non-revenue passes to go home, because Lee was hoping we would get the First Class cabin. The Air France flight attendant met us as we boarded, and said,

"Bonjour, Madame, may I welcome you to my First Class cabin. "

We started to enter the cabin, and then he stopped us.

"Excuse me, but do you prefer French Fries or Freedom Fries my friends?"

"We absolutely prefer *French* fries," I said with my sweetest smile.

"Please, you and your husband join the others for wine in the First Class Cabin."

We had a wonderful flight to Paris with Pierre, and as we were leaving he placed two bottles of French wine in my bag.

"Merci," I said as we began moving toward the exit with the other passengers.

"Now if we can just make our flight home with no problems, I will breathe a sigh of relief," I said to Lee as we entered the enormous, crowded terminal of Charles DeGaulle Airport.

We had about two hours before our Delta flight which would depart, and so we went to a restaurant and bar area to get something to eat and drink . I was listening carefully to the announcements which were in French, but then noticed that Delta was making announcements in both French and English.

We started for our gate area, and then found that the airport was so large that once through security, a large two story bus pulled up to a loading dock to take us to the gate of the aircraft which we would be boarding. I made it through security with no problems and boarded the bus, but suddenly I heard alarms ringing, and when I looked back Lee was surrounded by airport security personnel, and others were running toward him. They would not allow me to get off the bus and go back, and so all I could do was watch as Lee handed his shoes to the security guard, and it appeared the entire concourse of the airport had shut down. More alarms were ringing, security guards had Lee surrounded, and I was horrified.

I tried to explain to the driver and the attendants on the shuttle I was on that I could not leave without my husband, but no one was listening. The shuttle pulled up to a loading area and I entered the luxurious Business class cabin of the Delta 777 which resembled something from a James Bond movie. This was the first time I had been aboard Delta's most beautiful aircraft, and it is certainly impressive to say the least. As I was taking in the large pod-like seating areas I explained to the flight attendant what was going on , and she seemed quite concerned. She offered to try and find out what was happening.

We were only about twenty minutes from the scheduled takeoff time, and I had been a flight attendant long enough to know that an aircraft does not wait for one passenger; the crew is not allowed to make that kind of decision. However, to my surprise, thirty-five minutes later we were still sitting on the tarmac, and

Lee had not shown up. Suddenly, he burst through the door, and the flight attendant helped him to his seat.

"What happened?" we both asked him at the same time.

"My damn Cole Hahn shoes that I had re-soled before we left had two strings hanging from them. They thought I had explosives in my shoes." He was obviously very shaken even though he was trying to act as though shutting down Charles DeGaulle's Delta terminal for several minutes was normal.

"Oh, my God," I said. "Is everything ok, now?"

"It must be or they wouldn't have let him leave," the flight attendant laughed nervously.

Within the next five minutes the door was closed, and we were underway for the most luxurious flight I have ever taken. We had steak and lobster for dinner. There were coffee carts, ice cream carts, and at some time during the flight they brought cheese and nut assortments around. The seats made into full-length beds, and each seat had an entertainment center with movies, games, music, and a GPS interactive map for tracking the progress of the flight.

Once again, I found myself wishing I had taken the plunge, resigned from ASA and applied to Delta, which I consider to be the finest airline in the world. I will believe until this day that the crew of that particular flight stalled as long as possible to avoid leaving my husband behind. They knew we were part of the airline family, and on that very unfriendly day in Europe for Americans, they made sure we got out safely.

Margo Deal Anderson

## 17 Graduation

No one came but my daughter. I was so tired and felt so unattractive on graduation day from flight attendant training, but I was also proud of what I had accomplished. I think I was just as proud of the wings as I had been of my Master's degree, if not more so.

I had offered non-revenue tickets to fly to Atlanta to all of my family: my dad, my mom, my brother, and to Lee, but my daughter who was about to graduate from the University of Florida was the only one who came, and she would be the one to pin the wings on when my name was called. She was very proud, but she was also very excited about having flight privileges to travel.

There were only nineteen left of us out of the over fifty students who started out in the "C" or third class of the year; we had made it through the month of March, through horrible flying weather, living in the Embassy Suites hotel, and passing a test every single day with a 90 or higher in order to remain for six weeks.

We posed for photographs by the fountain in the lobby, and my daughter, Hilary, took photos and hugged me, telling me how proud she was that I had made it through.

A few minutes later she went inside to take her seat and the rest of us filed in as a class to begin the ceremony. As I listened to the comments of our instructors and waited for the "pinning" part of the ceremony to begin when family members would pin the wings on the graduates, I looked at the faces of each of my colleagues and thought of what we had been through together.

Six weeks ago we didn't even know each other, and now we were all very close, as though we had been in school together for years. I cannot explain the bonding that takes place in a class of flight attendants, but the training is so different than anything most people can imagine, that there is almost a survival instinct that sets in among the participants; there is not really any competition, just everyone trying to help each other make it through. Every time someone left the class the rest of us were saddened and became more paranoid.

The dress code for training included navy or khaki slacks or skirts, tuck in button-down white shirts, a belt, and either navy blue or black shoes. Hair had to be above the collar, nails had to be manicured, and makeup was required for the females.

The first test we had to take included memorizing the airport codes for all the cities Delta Connection would be flying, and this list would grow in the coming weeks as ASA was completely purchased by Delta and the two airlines merged. This merger was actually wonderful for flight attendants because the non-revenue flying benefits expanded and ASA flight attendants were able to fly to any destination that Delta employees could fly. This was another reason I found it so difficult to leave ASA and start all over again in seniority and training to be a Delta flight attendant...the benefits were exactly the same.

As I looked at the faces of the others, I also thought of those who did not complete the training because they found out they really hated flying, were afraid to fly, or became airsick when flying. After each exhausting week of training, on Friday we did not get the weekend off; instead, we lined up at the supervisors office and were given tickets to board ASA aircraft and we spent the entire weekend flying to as many destinations as they could schedule us for. Our job was to watch the working flight attendants in action, making their announcements, learning how to do the safety checks before the passengers came aboard, helping with serving and cleanup, and gradually, as the weeks went by, we actually did most of the work aboard the flights. The only thing we were not allowed to do was to sit in the flight attendant jump seat; one had to be a certified flight attendant to sit there because in the event of an emergency evacuation of the aircraft it is the primary duty of the flight attendants to evacuate the passengers, and the main exits are beside the jump seats.

I sadly thought of Russell, my favorite guy flight attendant who had been fired and not allowed to participate in graduation because he showed up wearing a female flight attendant uniform and a wig; he was a female impersonator, and I was sad to see him go. He actually looked better in the dress than any of the other flight attendants I knew.

Then there was, Peggy, the divorced doctor's wife who was fired when she received her first call from crew scheduling and told them she could not make it because she had a nail appointment. Crew scheduling had shouted so loudly at her on the phone that all of us in the crash pad could hear him telling her "to get her ass to the airport or she would be without a job." Peggy promptly told him that she had never "heard anyone so rude in her life, and that she didn't like ASA anyway." And with that, she walked away from six weeks of flight attendant boot camp hell and never looked back. Many of the others who were in our class in the beginning were just a blur; my roommate, who was a beautiful African –American girl passed all of the training and the final exam, but decided not to graduate or begin reserve. She was dating an Atlanta Falcon, and he didn't want her flying. My thoughts were racing as I saw Elizabeth, the Italian-American in our class, slightly plump, with perfect makeup, hair, and nails, who cried during every test until one of the instructors would sit beside her desk and practically answer the questions for her, begin to approach the stage for her mother and father to pin the wings on her vest.

"Margo Deal," I heard my name called and I stood up to go forward and receive the wings. I was still using my unmarried name in March of 2000 because Lee and I did not marry until December of 2001. My daughter stood up when I did and pinned the wings on my navy blue vest before giving me a hug. I had done it! I was a flight attendant!

Hilary had flown to Atlanta from Gainesville where she was

enrolled in pre-law at the University of Florida, and she spent the night with me at our crash pad in Stockbridge, Georgia before returning the next morning to Gainesville . She was super-excited about having the pass privileges for flying, and I was excited for her as well. I felt like this was something wonderful I had been able to do for her finally, after so many years of feeling rather inadequate as a single mom when it came to being able to afford nice vacations or cruises as the families of many of her friends were able to do.

As I drove Hilary to the airport the next morning, I was sad for her to be leaving, and I got out of the car to hug her before watching her disappear into the crowd of people hurrying inside the Hartsfield International Airport.

When I arrived back at our three-bedroom apartment in Stockbridge, my roommate, Valerie was in her bathing suit headed to the pool.

"O.K., let's get some sun and relax a little before the damn beeper starts going off and we have to fly somewhere, " she said grimly. "I feel like this is a calm before the storm."

"I feel scared," I said. "There are so many things that can go wrong. I would hate to get fired on my first trip."

"You won't," Valerie said matter-of-factly. "You only get fired if you don't hear your beeper and don't show up, of if you get caught in traffic and it takes you more than an hour to get to the airport."

"That's wonderful," I answered sarcastically, "I feel so much better now." I headed for our apartment to get my swimsuit. "I'll be right back."

Just as I turned for the apartment, Valerie screamed. I looked back and she was holding her beeper up in the air. I watched, horrified, as she pulled her cellphone out of her bag and dialed. She then threw the phone back in her bag, grabbed her towel off the chair, and walked quickly in my direction.

"The Hell has begun," she declared. "I am assigned to a three-day trip on the Brasilia; I'm going to Texas."

We both stared at each other for a second and started laughing hysterically as we ran toward our apartment together.

## 18   Coach and the Federal Judge

"I am blind.  There is nothing wrong with my ability to walk, to have intelligent conversation or to count the number of seats to the emergency exit," said the distinguished silver-haired gentleman with a beautiful black lab service dog.

"I'm sorry, sir, I didn't mean to offend you," I apologized as I moved my hand off his sleeve, and said politely," I was just following our procedures for boarding visually–impaired passengers."

"No offense taken, young lady," replied the gentleman.  "Coach and I will be sitting in Row 2, Seat B."

We were flying on the ATR today, and so this meant we would be walking all the way up the aisle from the rear entrance. Apparently "Coach" was the beautiful black lab service dog's name.

I led the way to the front of the aircraft, this time without touching or speaking to the gentleman, and as he placed his hand on the back of each seat, it was obvious he was counting the rows, and this was definitely not his first rodeo. He reached to open the overhead compartment, and he began removing his long black London Fog raincoat. "You may assist me with this coat, if you don't mind, " he said.

"Of course, sir." I replied, and I reached to help him remove each arm from the coat, which he handed to me. He sat down, and Coach lay down in the small area by his feet, moving as far under the seat as possible. This was obviously not Coach's first flight either. I turned to walk back to the aft of the aircraft where I was the "A" flight attendant, but before I walked away, I said, "Please let me know if there is anything you need, sir."

"Thank you, young lady," he replied.

There were no other special assistance passengers for this flight to Asheville, North Carolina, and I quickly finished my safety checks and straightened the galley as I waited for the other passengers to arrive. Linda, the "B" flight attendant was still on reserve, and she was in the cockpit flirting with the First Officer, Brian. I didn't really mind; I was not in the mood for conversation with someone I didn't know today. I had a really bad sinus headache, and I didn't want to fly because I knew it was going to hurt my ears. There is no such thing as sick leave when you are a flight attendant for ASA; you receive an occurrence when you are sick, and if you get too many occurrences you lose your job. I had taken a Sudafed, hoping it would help, and I had also used Afrin spray, the staple over the counter drug used by all flight attendants, but nothing was helping to relieve the pressure.

I saw the passengers coming from the terminal, their heads bent to shield their faces from the cold wind, and there was a slight misting rain which was chilling straight to the bone. Atlanta can be brutally hot in the summer months, but nothing can be colder than a misting rain when the temperature is just above freezing in the middle of winter in northern Georgia. The locals call it a "wet" cold, and it literally makes your bones ache. All I could think about on days like this was a hot shower when we finally got to our hotel after a brutal day of loading and unloading the ATR as we traveled between Atlanta and Augusta, Tri-Cities, Asheville, and Wilmington.

Once the passengers were on board, I went to the cockpit to remind Linda that she needed to do a passenger count before we took off, and I noticed that she had her hands on Brian's shoulders as she spoke with him; I felt like her mother reminding her to do chores. I turned and walked back through the baggage loaded on either side of the hallway leading to the ATR cabin from the cockpit, and as I looked down at Row 1, I saw the large paws of Coach protruding from underneath the seat. I smiled and thought, *"What a great dog."*

The paperwork was finally ready, and after I handed it off to the ramp guy, Linda went forward and I saw that she was ready with the safety demo equipment. I waved to the ramp guy as the door was closing, and just as I was ready to begin the announcements, the flight attendant call button sounded from Row 8. I quickly walked forward to see what was wrong, and the passenger, with a worried look on his face asked, "Why is only one of the props turning? What's wrong with the other one?"

I smiled to myself, and although I wanted to give him the answer and tell him that the Captain only uses one prop for taxi out to save on fuel, but flight attendants are not allowed to answer technical questions according to ASA policy. Instead I said, "Just a moment sir, and I will ask the Captain and let you know."

I picked up the phone and said, "Just checking to see why we only have one prop working," I laughed. "God, this question comes up a hundred times a week; can I please just answer without letting them see me pick up the phone and ask the crew?" The Captain laughed and said, "I tell you what. Count to ten, walk over and kick the side of the wall just under where the prop is."

I started laughing just thinking about it, and said, "ok."

I walked over after counting to ten, kicked the wall, and the other prop started turning, right on cue.

"Darn, it's been doing that all week," I said angrily as I turned quickly and went back to the phone to begin the safety announcements, leaving the passenger in Row 8 with an astounded look on his face.

This was one of my best moments in flying on the ATR ever, but it was short-lived. As soon as we took off my ears began to feel as though someone was compressing my head in one of those medieval torture devices. I covered my ears with my hands and just prayed for relief, but none came. Although it can certainly fly at higher altitudes, the ATR normally flies at an altitude of 14,000 to about 18,000 feet, and for some reason this is the area that seems to hurt the most. The pain was so great that when Linda saw me holding my ears, she knew what was happening: she offered to serve the cabin by herself. The flight to Asheville takes about an hour and a half, and we only had forty-five passengers aboard which gave her plenty of time to serve. I thanked her profusely, and I felt badly about thinking that she was going to be a lazy co-worker when I saw how much time she was spending in the cockpit back in Atlanta.

When she finished serving, she told me that the distinguished blind gentleman in the front was a Federal judge and that he flew on this particular flight twice per week.

"Really," I said. "He seemed to know what he was doing, and you can tell the dog has flown a lot."

"How are your ears?"

"They hurt like hell, especially the left one. I have had this problem since last winter when we were flying out of Dothan, Alabama. The cabin had a pressurization problem, and I cried for over an hour until we got off the plane. I went immediately to the emergency room and they sent me to a specialist; I had blood in both ears, and I was not allowed to fly for sixty days. I have had problems ever since, especially if I have a cold or any sinus congestion like today. That's what makes me so mad; it's the airline's fault that I have a problem, but they won't allow sick leave without giving me an occurrence. I hate ASA; no other business in the world treats their employees like this."

"I know," Linda agreed. "I was talking to a Delta flight attendant, and being on reserve for Delta is nothing like ASA; I absolutely loathe those guys in crew scheduling. I have flown almost ninety hours already this month, and I know other reserve flight attendants that haven't even been called at all yet. I am sure they have their favorites who get special treatment, and I am definitely not one of those. One of the girls in our crash pad told them she was on her period and had cramps, and they didn't make her take the trip. How is that fair?"

The four bells signaling initial approach into Asheville rang, and I picked up the phone to ask the passengers to raise their tray tables and seat backs, put away portable electronic devices, and to make sure their carry-ons were completely stowed under the seat in front of them.

Linda asked if I was ok, and I lied and nodded yes. I forced myself to get up out of the jump seat, and I walked down the aisle with the Delta "thank you" bag collecting trash and checking to see that everything was stowed. The air was bumpy as always over the mountains, and I noticed Coach's large black paws stretching out from under the seat to give him more grip on the carpeted floor. I smiled as I walked past him, resisting the urge to pet him which was a "no-no" with service animals.

The pain had either lessened in my ear, the Sudafed was kicking in, or I was just numb because it had hurt for so long; at any rate, by the time I fastened my seat belt and announced to the passengers that we were on final approach and would soon be on the ground in Asheville, I decided that I might live. I even managed to smile as the passengers were deplaning, and I summoned enough courage to ask the Judge and Coach if they would like any assistance with the stairs.

"Thank you," he said, "but I think I can manage. Is the gate agent waiting at the bottom?" he asked.

"Yes," I said, "standing on the right-hand side."

" I will see you on Tuesday for my return flight," he said.

"Take care," I replied. I decided not to bother to go into the whole description of what my schedule was and was not. I would not be on his flight on Tuesday, but I would definitely probably see him again if he flew often with ASA, which he obviously did. Many passengers think that flight crews fly together all of the time, flying the same schedule back and forth from one city to Atlanta, when in reality, most of the time flight crews have only just met when they enter the aircraft at the beginning of a trip. Because ASA was a fairly small airline in comparison to the others, when I first began flying, it was not unusual to fly with crew members I had flown with before, or who I knew from attending training together. Also, because the bidding process for monthly trips was done according to seniority, the same crew members were sometimes awarded the same bid because of their placement on the seniority list.

After the passengers had finished deplaning, we went back through the cabin picking up trash that had been left behind, folding blankets and pillows and putting them in the overhead compartments, and finally crossing the seatbelts on the seats so that passengers didn't have to reach under the person's butt sitting next to them searching for the ends of the seat belt. Most airlines do not even have pillows and blankets in the coach section anymore, especially for shorter domestic flights, and I always cringed when I saw passengers using blankets which had been used on previous flights; clean, unused blankets were not placed on the flights until the end of a day when they returned to Atlanta. or sometimes overnight stations provided clean blankets. Otherwise, the passengers got the same ones that had been sneezed on, coughed on, or breathed on by the passengers on previous flights. I did have a stack of paper pillowcases in the galley which I tried to put on pillows when I had time in between legs; as much as I disliked ASA, I really did try to follow the guidelines of treating every passenger as though they were a member of my family. I wouldn't have wanted my mother placing a pillow under her head with a dirty pillowcase on it.

We had two "legs" or flights left on today's schedule; we were taking passengers from Asheville to Atlanta, and then we would be flying from Atlanta to Augusta, Georgia, one of my favorite overnights, in the beautiful old Partridge Inn. The Partridge is a boutique hotel which is used during the Masters Golf Tournament in Augusta each Spring, and we would have an eighteen hour layover there before returning to Atlanta. I never understood the staging of the trips, when sometimes we would barely have eight hours to sleep after a ridiculous day of six or seven legs, and then other times we would have up to twenty-four hours to lounge by the hotel pool, sleep, watch television, and go out to eat, but that is the life of a flight crew member.

I stepped into the lav and used my Afrin nasal spray before the passengers began boarding for Atlanta, and I immediately felt my left ear clear when I gently blew my nose.

"Thank you, God," I whispered as I stepped into position to greet the passengers as they boarded the flight for Atlanta.

## 19   Non-Revenue Flying

The best part of being a flight attendant is the ability to go anywhere , anytime.  A great deal of satisfaction can be derived from knowing that one can go to Paris on a Friday night, fly in the first class or business elite cabin, and then return to Atlanta on Sunday as though dinner in France is a normal occurrence. I remember one time when I first started as a flight attendant, one of the senior flight attendants told me that she had been so broke one month that she could not afford a hotel and was stuck in Atlanta, and so she booked herself in first class and flew to London and back, watching movies, drinking wine, and relaxing instead of being stuck in the flight attendant lounge for two days in a recliner eating out of the vending machine.

I was so excited after flying for thirty days when I was able to offer flight benefits to my family; the way it works is that the spouse, parents and children receive the same benefits as the employee, and if the employee is not married, he or she is allowed to designate a "significant other" who also enjoys the same flight benefits. The "significant other" may be a boyfriend or girlfriend, a fiancé, a brother or sister, or just a good friend. Additionally, each employee also gets eight "buddy passes" per year which can be given to anyone. The buddy passes are not free like the family passes; the passes basically charge pennies per mile plus taxes. For example, a ticket from Panama City , Florida to Atlanta, Georgia might cost around $40.00 using a buddy pass. It's a pretty good deal, but the seating is space available, and the priority is determined by the seniority of the employee. In other words, if you have been employed since 1978, you are going to get on the airplane before someone who has been employed since 2014, and this applies to your family members and buddies as well. Active-duty employees are given a higher priority than retirees when flying, and there is a saying in the airline industry that the quickest way to lose a friend is to give them a buddy pass.

One of the best perks to non-revenue flying is getting to fly in the first or business class seats when they are available. When I first started flying employees were not allowed to wear denim (as in blue jeans) and fly in first class, but that has now been relaxed. Right after 9/11 when almost no one was flying for about six months, non-revenue flying was the best ever. For those who were brave enough to fly, the first class sections, as well as the rest of the seats on almost every aircraft were wide open to almost any destination. It was definitely the time for employees to take great vacations and not worry about getting a seat.

My husband started flying with me on many of my trips because he was able to get a seat so easily; we would check the computer to make sure he could get on every flight, and then at the end of the day he stayed in my hotel room with me.  Sometimes, if I had a particularly long day with five or six legs of flying, he would just go ahead and fly directly to where I would be staying that night and enjoy the amenities of the hotel, lying by the pool all day, watching television, and sometimes sightseeing.

We explored the San Antonio Riverwalk, barhopped in Jacksonville along the St. Johns River, drank margaritas in Monterrey, enjoyed the boardwalk at Myrtle Beach, and went sight-seeing in Asheville, all the while enjoying free lodging at luxury hotels at airline expense.

We spent a frozen Christmas Eve together once in Toronto, and had a blast partying with others who were stranded at the beautiful Renaissance Toronto Downtown Hotel, a luxury modern hotel built into a sports stadium.  The Toronto Blue Jays play there, and the stadium roof will actually roll back revealing the beautiful Canadian sky.  Flight crews loved staying at this hotel during baseball season, because the rooms had windows looking out onto the baseball field!

One of the most romantic New Year's Eve nights we ever spent together was when our flight was snowed-in for "First Night" in Worcester, Massachusetts.  The entire crew was stranded there for four days because of a snow storm;  my husband and I crashed a wedding party, watched fireworks, built a snowman in the middle of the deserted streets, and spent hours in bed watching movies in the beautiful Hyatt Regency.

The hotels provided for flight crews were for the most part, really nice, and each crew member is provided a private room, something I was extremely thankful for.  After standing twelve or more hours , loading and unloading passengers, serving, cleaning up, making announcements, and dragging luggage all day, no one wants to share a room, especially a bathroom with someone you either don't know or just met.

Spending time together while I was working was great, but during my days off, Lee and I used the flight benefits for employees as much as possible. During the five years I was employed by ASA, I probably logged as many non-revenue hours as I did work hours, maybe more. Germany, France, England, Belgium, Costa Rica (several times), Cayman Islands, Aruba, Hawaii, St. Thomas, Puerto Rico, Acapulco, Cancun, Cozumel, New York, Chicago, Alaska, Washington D.C., Los Angeles, San Francisco, Minneapolis (for Christmas shopping), Cleveland, Indianapolis, Boston, San Antonio, Dallas, Houston, Phoenix, Scottsdale, Belgium, Miami, Key West, Mexico City, Cabo San Lucas, Montreal, Toronto, Monterrey, and St. Martin, are all places we traveled using not only the flight passes, but also the hotel and resort discounts.

Acapulco was one of the most memorable places for me; we stayed at Los Flamingos, a hotel made famous by the Hollywood gang, (John Wayne, Erol Flynn, Deborah Kerr and others), and owned by Johnny Weismuller, the Tarzan who was loved by the people of Acapulco so much that to this day the hedges around the beautiful pink cliff side hotel are trimmed in the shapes of monkeys and alligators. The famous coco-loco drinks are served in coconut shells in a bar with a sheer drop-off of about 1,000 feet straight down where the Pacific Ocean crashes against the rocks. We ate a candlelight dinner at the El Mirador Hotel and watched the famous cliff divers who pray before the Virgin Mary before diving into the waves and then climbing up to the dining room with palms outstretched for tips.

As much as I disliked so many of ASA's regulations, working conditions, and the stigma attached to working for the commuter airline with its "puddle jumpers" and "Barbie jets," the non-revenue flying was the benefit which motivated me to remain a flight attendant for five years. Being able to fly anywhere in the world that Delta flies, as well as with her partner airlines, is an amazing experience, a feeling of complete freedom to go anywhere, anytime.

Margo Deal Anderson

## 20  Recurrent Training

"I heard almost everyone failed; only two people came back to Dallas last week from recurrent," said Elizabeth.

"How is that possible?" I asked. "Recurrent isn't hard; it's just a pain in the ass."

"Joseph said that the instructors were really tough; he said there was absolutely no slack about anything," answered Elizabeth.

"Great," I said glumly. "So if we don't pass we have to stay in Atlanta and do it again?"

"Yep."

I settled into the seat of the MD88 as we headed east to engage in what had to be the most aggravating part of a flight attendant's job---recurrent training.

Once each year, flight attendants have to attend a refresher course to make sure everyone still knows how to evacuate an aircraft, how to do safety checks, perform CPR, operate a defibrillator, and an assortment of other information from hospitality and marketing skills, and now the new addition to our training, self-defense from terrorists, thanks to the 9/11 attacks.

I was now based in Dallas, Texas instead of Atlanta, Georgia because Panama City started non-stop jet service to Dallas, and it was actually easier to commute to Dallas than it was to Atlanta. An added perk was that my seniority would be higher in Dallas since it was a smaller base, and I would be able to fly part-time rather than full-time. I loved this, because by 2005, I was really tired of flying and living out of a suitcase, but most of all I was just completely sick of ASA.

Lee and I had been married almost four years, we had purchased a beautiful little chalet on a lake near Panama City, and I wanted to be home more than I was. When flying part time, I was able to bid on the schedule I wanted for the month, pick half of the trips, and then the other half of the trips would be given to reserve flight attendants or picked up by flight attendants who wanted to make extra money. This meant I was flying about 35 hours per month and still able to keep the flight benefits for both Lee and I; during the last months of my employment with ASA we actually logged more miles flying for pleasure than I did flying for work. I did not feel guilty about this at all because during my first three years with ASA I was extended so many times by crew scheduling on trips that I more than paid for any perks during the last months of my job.

I had flown to Dallas to "duty-in" which is the flight crew term for clocking in, and then immediately, four of us went to the terminal to board our flight for Atlanta and four days of recurrent training. The scary part about recurrent is that if for some reason you do not pass, you have to remain in Atlanta and repeat the recurrent before you can fly again. I just wanted to get there, get it done, and go home.

The first day was not too bad, just a review of safety and first aid, or "first care" as it was termed in our training. This type of emergency medical treatment which would be for passengers instills in flight attendants that the most important thing is to save the life and to "do no harm" while performing life-saving actions. We were told that flight crew members are held harmless in the event that someone does actually pass away while we are attempting to help them, because at 35,000 feet, you are known as a good Samaritan, treating a passenger in harm's way the same way you would want someone to treat one of your family members in the same situation. With all of that said, the best first action to take is to find out if there is a doctor, nurse, or EMT aboard the aircraft before undertaking any first aid or life-saving procedures.

The second day of recurrent was the worst; we had to get in a simulator cabin that filled with smoke and had lightning and thunder noises on the outside and shout out the emergency evacuation commands while evacuating the "passengers" from the aircraft with the recurrent training supervisors holding a timer. The "passengers" were captains and first officers who were just as horrified by our shouting as we were, and I got so confused the first time that I shouted, "Follow me, follow me," which is not the proper command. This would insinuate that I was the first one out, asking passengers to follow me instead of my remaining behind to assist them out.

We went through dress code, attendance policies, sick leave policies, and international travel procedures. The written test was much harder than what we had been given in the past, and I was quite sure I had failed it even though I am an ace at taking almost any kind of test that involves a lot of memorization, which this one usually does. Elizabeth, one of my comrades in misery from Dallas started crying right in the middle of the test and said, "I hate this f—king airline," before running out the door to the bathroom. I started to get up to go and check on her, and the instructor told me if I got up to leave I was considered finished with my test.

"Fine," I answered sarcastically as I sat back down.

"You don't have to finish it at all if you would rather just come back," said the instructor.

"No problem, I'm finishing." " *Good thing you aren't a mind-reader*," I thought to myself.

Thirty minutes later I was waiting with the group of about thirty flight attendants to find out if we passed or not. The four of us from Dallas huddled miserably together, and when the list was posted I passed and one of the others passed; two of the Dallas group, or 50%, did not pass and had to stay in Atlanta.

"Jesus," said Mary the other flight attendant who passed. "Let's change out of uniform and get a drink before we go to the gate."

"Great idea; you don't have to ask me twice," I agreed.

We went to the flight attendant lounge rest room and changed into jeans, then headed straight for the Atlanta Chili's and Chambord raspberry margaritas in fish bowl glasses. After two of them we headed for our gate to use our "1" designation non-revenue seats which means the same thing as confirmed seating; using this designation to fly standby any other time without authorization can result in charges against your paycheck for $150.00 or more, and in some cases the revocation of flight privileges.

By the time we boarded the flight we were both feeling a little "buzzed" and probably talking too loud.  As we took our seats in the first class section of the airplane we were probably even louder, but we were incredibly relieved to be finished with recurrent training and also to be headed back to Dallas, and then home.

"Wherever you two are going I want to go," a voice from behind us said jovially.

Both of us stood up on our knees and looked over the top of our seats at the gentleman seated behind us.

"A drink for us and one for our friend," Mary shouted to the flight attendant who was serving drinks to the others in first class.

The flight attendant walked over to us and said in a quiet, meaningful tone, "I don't know what you two have been drinking, but if you don't tone it down you are going to be staying in Atlanta."

"Sorry," said Mary. " We are just very happy and  relieved after passing recurrent."

We both sat  down and faced the front, and then broke into silly giggles.

The flight attendant finished serving the row across from us and then bent down and said, "I almost hate to ask you this, but would you ladies like something to drink?"

"If we promise to be very quiet, may we both have a cocktail?"

"Certainly," she said. "My pleasure."

"I'll have a vodka and cranberry."

"Ditto," said Mary.

"Dallas, here we come," I said to no one in particular, knowing that there would not be a flight home to Panama City until tomorrow morning, and I would be spending the night near the airport in the world-famous Motel Six, Irving, Texas, a cheap, clean place to sleep that caters almost exclusively to the dozens of flight crews who are in and out of there almost daily waiting for trips to begin or for a flight home.

## 21 Embraer Brasilia Hellcraft

I had been so depressed after we received our bids for the month and I saw that I had been assigned to the Embraer EMB 120, a twin-turboprop commuter airliner which seats 28 people and in which I cannot stand up straight because I am 5'10". I had been on a trip when our bids were due, and since at that time ASA was not using computer bidding yet, my bid was late and so I was assigned what was left.

Today we were headed for Columbus, Georgia, just a twenty minute flight away from Atlanta, and the pilot I was flying with was an ex-Marine, known for narrating and chatting with passengers the entire flight. I had flown with him one other time, and I was ready to jump before we finished the trip. This month's bid package included stops in every small town across Georgia and Louisiana, and I knew it would be absolutely miserable.

ASA/Delta Connection no longer uses the Brasilia or the ATR aircraft, but during the five years I was flying with them they were the workhorses of the airline, and while I hated the ATR, I was actually afraid of the Brasilia because of its crash record.

On April 5, 1991, Atlantic Southeast Airlines Flight 2311 crashed at Brunswick, Georgia. The crash claimed the lives of twenty-three people on board, including former U.S. Senator John Tower of Texas and astronaut Sonny Carter. This was due to propeller control failure; and then on August 21, 1995, ASA Flight 529, crashed in a field near Carrollton, Georgia. Of the twenty-nine people on board, ten died (one casualty was from a heart attack nearly 8 weeks later). This was due to failure of a propeller blade and subsequent severe engine vibration and failure. There have been many other crashes of this aircraft, but these were the only two operated by ASA/Delta Connection: two were enough for me to have something to think about every time I was assigned to the aircraft, however.

The flight attendant seat was just inside the passenger door, and on the other side of the passenger seat was the lav, or the "head" as most flight crew members with military experience call it. The galley is just in front of the flight attendant seat and there is a row of passenger seats directly across from the galley, leaving the butt of the flight attendant directly in the face of those passengers, especially when bending and stooping to prepare drinks and snacks. The beverage and snack service is done using a tray on this aircraft because it is too small for a cart, and I had to wear flat shoes and stoop as I moved about because 5'10" is absolutely too tall to be working on this horrible little aircraft.

The only bright spot is the last row which has five seats all the way across the back; if one is lucky enough to fly on the Brasilia when for whatever reason there are only a few passengers, it makes a great place to sleep.

Today, as I pressed my back against the flimsy little folding door leading to the cockpit, I made myself as small as possible as the passengers began boarding, stepping on my feet, hitting their heads on the doorway, and complaining about how small the aircraft was.

"Damn, puddle jumpers," a large man exclaimed as he simultaneously bumped his head on the door and tripped as he started up the aisle to his seat.

"Can't you make these planes a little bit smaller," he sneered as he flopped unceremoniously into the third row.

"We try," I said, "but the company in Brazil said this is as small as they can make them."

"Oh great," he replied. "Now we have a comedian for a flight attendant."

"I'm sorry," I laughed. "I was just kidding. Would a complimentary drink help at all?"

"Your damn right it would; they should make the tickets complimentary just for flying on this piece of shit," he said.

My face turned redder than it already was in the hot, humid, Atlanta heat, but I tried to sound pleasant as I said, "I will get something for you as soon as we get in the air."

I then turned quickly back to greeting the other passengers as they climbed the steps from the tar mac.

I had all male passengers today, and most were wearing business suits, carrying their jackets and displaying large wet spots on the backs of their shirts and in the armpits, thanks to standing on the tarmac for the last ten minutes as the gate agent patiently explained to each one why their carry-on baggage was not going to fit in our overhead bins and how the pink valet tags would assure that their luggage would not be lost or go to baggage claim, but rather, would be handed to them at plane side at their destination.

Most wore disgusted looks on their faces as they entered the plane, and after everyone was seated, my passenger count was at seventeen.

*"Well, maybe today won't be too bad after all,"* I thought to myself. *"Twenty minutes to Columbus, a twelve-hour layover to relax at the Sheraton , and then off to puddle-jumping through Louisiana and Texas for the next three days—God."*

The first officer had already completed his walk-around outside the aircraft, and he was working on baggage and passenger count paperwork as Tom, the Captain entered. He was large, red-faced, and actually smiling as he stood directly in front of me and addressed the passengers, almost pushing me aside into the first row of seats in the process.

"How's everybody doing today? We are going to have a short but bumpy ride over to Columbus, so make sure your seat belts are fastened. The clouds are starting to move in for some afternoon, regularly scheduled thunderstorms. "

He turned to enter the cockpit, and I had gotten back up off the front seat where I had sat down to get out of his way. I had flown with him before, and while I sometimes became annoyed with his announcements and loud voice, he was actually a pretty nice person. I made up my mind to try and be as congenial as possible.

I made my announcement, demonstrated the safety equipment, and within just a few minutes the paperwork was handed out the door to the ramp guy, I was pulling the door closed, and I sat down in the tiny flight attendant jump seat staring into the faces of seventeen hostile business men. We were not allowed to read magazines or books in front of the passengers as part of ASA policy because we were expected to smile and carry on friendly conversation. We *were* allowed to look at our flight attendant manuals, and so I pulled it from my bag in the overhead and began to look through the pages as though I were looking for important information in order to avoid looking at the men.

"Why are you reading the manual, honey," the gentleman in Row Four quipped, "you don't need to know how to fly the plane." There was a round of appreciative laughter.

"You are certainly right about that," I answered. "I just wanted to make sure I knew how to open the door correctly in case something goes wrong and we have to get out." The laughter stopped.

I would never have said something like this to anyone, but I really get tired of being treated like a brainless blonde sometimes, and today it just pissed me off, especially when the other passengers laughed.

Well, they weren't laughing now, and it was almost silent on the airplane except for the loud sound of the turbo-props as they came to life and we began to move. I continued to stare intently at the manual, and out of the corner of my eye, I could see that the passengers continued to stare intently at me. I felt as though I might get the silly giggles, but I restrained myself, and finally we were in the air and I could get up and start the beverage service.

The flight had been late due to maintenance, and so I offered the passengers a complimentary adult beverage, and this certainly lightened the mood a bit. I don't know why all commuter airlines don't just hand everyone a beer as they get on the plane. It would certainly make things easier for the flight attendants.

The liquor kit on the Brasilia is tiny, with only a few choices, but two Jack Daniels, three Crown Royals, four Smirnoffs, two red wines, one Dewars, three beers, and two coffees later, things were much less tense and I called the cockpit to see if they wanted something.

Just as I picked up the phone to ask, the Captain interrupted the silence with a booming voice over the intercom, "Good afternoon gentlemen, and welcome to Atlantic Southeast Airlines, Delta Connection's service to Columbus, Georgia.."

"*Jesus,*" I thought to myself, "*Here we go. You would think we were on an international flight.*"

Tom's announcement went on and on, describing altitude, weather conditions, the view from the left, the view from the right, the history of the Embraer Brasilia aircraft, the time of the flight, and our estimated arrival time. As he finally finished, I heard one of the passengers mumble under his breath, "Son of a bitch, does he ever shut up?"

I laughed out loud. Then, I picked the phone up and asked the Captain, and his First Officer, Paul, if they wanted anything to drink. They both said no.

"*Good,*" I thought. "*I'm done until we get there.*"

Two minutes later the Captain came back on the PA system and announced, "Please make sure your seatbelts are securely fastened; we have some rough air ahead."

Just as he made the announcement the aircraft began to bump and then we were in some *serious* turbulence. We dropped enough one time that it really scared me, and I tightened my seat belt. Then we settled into sort of a rhythm of turbulence which was extremely rough, and when we dropped suddenly again, one of the Jack Daniels passengers yelled out, "Yee…..Haw…." as he raised his glass to the top of the aircraft to keep it from spilling.

Then the others got into the spirit and they all held their drinks up and began to join in with, "Yee Haw, Ride'em Cowboy, Woo Hoo!"

"*Oh, my God*," I thought, "*three more days of this, are you kidding me?*"

I heard the four tones which signaled initial approach, and I made the initial approach announcement, but no one could hear me because they were having so much fun pretending to be cowboys. Initial turned into final approach, and we made a surprisingly easy landing considering the weather. As obnoxious as he was, Tom was a good pilot, and his Marine background always made me feel as safe as I ever felt on this horrible little airplane.

"Welcome to Columbus, gentlemen. Thank you for choosing ASA, and we do hope that when your future plans include air travel that you fly with us again.  We wish you a wonderful stay here in Columbus, or wherever you final destination may be…….."

I grabbed my flight bag and followed the crew down the steps, into the terminal and out the door to our waiting shuttle which would take us to the Sheraton and twelve hours of peace *away* from airplanes and passengers.

## 22 TEXAS, BROWN WATER, COWBOYS, AND ST. ELMO'S FIRE

*"Oh God, I hate Texas,"* I thought as we left Atlanta for San Antonio. It always seemed that the hours we flew over Texas were longer than other flights; the state seemed to go on forever, and fifty passengers can work one flight attendant almost to death in that amount of time.

After I finished serving, I noticed lightning in the distance as I looked out of the small window in the door of the galley; weather always seemed to be extreme in Texas; the heat was oppressive in the summer, but in winter the cold wind at the Dallas/Ft. Worth airport can absolutely take your breath away. A few of the passengers asked for seconds on their drinks, but after I picked up trash and I walked up and down the aisle again to see if everyone was settled, most of them were either reading, working on laptops, or had turned off their lights and were going to sleep.

I went back to the galley in the front of the cabin, pulled my bag out of the closet, and sat down on it with my back against the galley wall where I could not be seen by my passengers. I sat there for a few minutes just catching my breath and relaxing, and just as I was getting ready to get up and make myself a cup of instant chicken soup and a Diet Coke, I heard the familiar tones indicating the cockpit was calling.

"What's up," I answered. "Would you like more coffee or another snack?"

"No," answered John, the First Officer. "We want to show you something."

"What?" I asked suspiciously. Flight crews were infamous for jokes on flight attendants, and I knew both John and Greg pretty well.

"Just come up here," laughed John.

"Ok," I said.

When I got to the cockpit, John wanted to know if I had ever heard of St. Elmo's fire? I had seen the movie, *St. Elmo's Fire,* but I had never really given any thought to what it meant.

"No, I guess I haven't," I admitted.

"Look out the window, and keep watching," said Greg, the

Captain.

As I looked, suddenly beautiful colors of light were arcing across the window and down the sides of the aircraft.

"Oh, my God, " I exclaimed, "that's beautiful!"

"St. Elmo's fire is a weather phenomenon --- luminous plasma is created by a corona discharge from a sharp or pointed object in a strong electric field," explained John. "In our case the object is the aircraft and the thunderstorms around us are generating the electric field."

"Sailors sometimes saw the light at sea on ships during thunderstorms, and St. Elmo is considered the patron saint of sailors, thus, St. Elmo's fire, " continued Greg.

Suddenly, more of the beautiful ghostly colors moved down the side of the windshield and disappeared.

"Thank you for calling me to come up here," I said. "I thought you were probably going to play a joke on me or something. This is beautiful."

In the days before 9/11 it was not unusual to visit the cockpit for a few minutes to take drinks and snacks to the crew, or sometimes just to chat, but flight attendants were discouraged from spending very much time away from the passengers, both for safety reasons, and just to avoid customer service complaints about the flight attendant not being available to passengers. Flight attendants had a key to the cockpit door in those days, but after 9/11, for safety reasons, the doors were reinforced, and we no longer had keys to open them.

"Well, I suppose I had better go and check on our passengers," I said as I reluctantly turned to leave the beautiful spectacle of St. Elmo's fire. "See you in San Antonio."

San Antonio's hotels were better than any of our other Texas destinations, and I always enjoyed the Riverwalk, having a great margarita, and looking for souvenirs in the cute little shops, but the rest of the destinations were not my favorites.

After leaving San Antonio the next morning, we would fly to Lubbock, then to Amarillo, and then back to Lubbock where we would spend the second night of our trip. I had spent the night in Lubbock before. The water in the Lubbock Holiday Inn was brown: brown in the toilets, brown coming out of the shower, and brown in the room service water goblets. It was disgusting. When we landed in Amarillo there was snow on the ground, and the cold air took my breath away as I opened the passenger door. I had never been to Amarillo, and all I could see was blowing snow and empty fields. I shivered and stood as far back from the door as possible as the passengers began deplaning and making their way across the tarmac to the terminal.

There were only fourteen passengers on the flight from Amarillo to Lubbock, and I made a fresh pot of coffee because they all looked blue from the cold. I decided to use a tray for serving since there were only a few passengers instead of dragging out the cart. Besides, as cold as it was outside the air would probably be bumpy and I did not like having the cart in the aisle when I was having to deal with turbulence. I wanted to be able to sit down quickly and buckle my seat belt. Flying through the air, possibly hitting the ceiling of the aircraft , or being injured by the beverage cart is a very scary and dangerous thing, and it happens more often than people realize to flight attendants who are serving beverages when unexpected turbulence occurs, or when the turbulence is rougher than the crew estimated it would be.

I took everyone's drink order and then went to the galley to prepare the drinks and snacks to serve. Almost everyone wanted coffee, and two of the passengers requested hot chocolate. The flight to Lubbock was so short that I barely had time to serve everyone and clean up before we arrived, and now we had a long layover at the hotel with the brown water. I stuffed as many bottles of water from the galley as I could fit in my bag, grabbed a package of biscotti cookies and some pretzels, and headed down the stairs. I also made sure I got my flight attendant manual from the cabinet because another crew was taking this aircraft, and we would get another one tomorrow morning. There is nothing more disconcerting that leaving something on an airplane (phone, ID's, purse, lunch) and realizing your belongings are now on their way to Atlanta or some other destination possibly never to be seen again.

We arrived at the Lubbock Radisson, and after checking in, going to my room and changing out of my uniform, I headed straight for the bar to get a margarita in a "to go" cup for my room. There must have been a ranchers convention going on because every man in the bar was wearing a cowboy hat, and the noise was deafening.

*"Wow, "* I thought, *"four o'clock in the afternoon and they are wide open in here."*

I made my way to the bar, ordered a margarita to go, and quickly left amidst a few cowboy comments I could do without. As soon as I returned to my room I headed straight for the shower to get rid of what I called, "airplane germs," because I always felt like I had been exposed to more than I wanted to know about, breathing the recirculating air on the plane as passengers coughed, hacked, and sneezed their way to their destinations. Just as I remembered, the water coming from the shower head had a bronze tint to it, and there were brown rings in the toilet and the sink.

"Ugh," I said out loud, as I hung my cosmetics bag on the back of the door and unzipped it to reveal my makeup, shampoo and conditioner, razor, cologne and deodorant, removed my toothpaste and toothbrush, and carefully placed everything on a hand towel which I spread on the counter next to the sink. I am just as paranoid about hotel germs as I am airplane germs.

Soon I was in my sweatpants, warm, over-sized sweatshirt, and a pair of Lee's thick socks, sipping a margarita and channel surfing for a good movie. We didn't have to duty-in at the airport until 11:00 a.m. tomorrow for the return flight to Atlanta; I called Lee to let him know I was safely "in" for the night, ordered room service tortilla soup, and another margarita. After watching "When Harry Met Sally," I fell asleep somewhere in the middle of "Out of Africa," and didn't awaken until the next morning when the alarm went off at 9:00 a.m.

## 23 Crew Scheduling

ASA's crew scheduling is an entity which is second only to the IRS in the ability to impart fear and loathing; if flight crews could get to them, they would all be subject to serious bodily harm. They are housed in an office in an undisclosed location, and they are ruthless, unfeeling, and uncaring. They are tricky, conniving, and unfair as they manipulate the lives of flight crews without any regard for illness, family problems, holidays or other special occasions.

The only job of crew scheduling employees is to make sure all of the aircrafts have flight crews; if you are a reserve flight attendant or pilot, crew scheduling can call you at any time unless you have "timed out" and are unable to fly until you have the required number of hours of sleep, and then they immediately call you again. While you are on reserve you can't just ignore your phone or beeper; if you do not call them back you get an occurrence, and when I was a flight attendant if you got three occurrences, you could lose your job. An occurrence stays on your record for a year before falling off, and a "trip failure," meaning you fail to show up for a trip stays on for two years. If you have two current trip failures on your record and receive a third one, you are automatically fired. You could not take a sick day until you had been employed for a year, and then only because of the Family Medical Leave Act, signed into effect by President Bill Clinton. There is no forgiveness from crew scheduling.

Pilots and flight attendants who were no longer on reserve with ASA, but had a regular schedule were also subject to the long arm of the crew scheduling employees; for example, once I had finished a grueling four-day drip on the ATR, and I had just sat down at the gate in Concourse C to wait for my flight home to Panama City. I was so happy because there were sixteen available seats, and I would be home in an less than two hours. My cell phone rang, and without looking to see who was calling, I answered it.

"Hello, Margo how are you this morning. This is crew scheduling, and we need you to do a round trip to White Plains for us."

"Oh, no," I wailed. "We won't be back until tonight, and there is no way I'll be able to get home until tomorrow. I start a three-day on Friday, and this will take almost all of the time I have at home with my husband this week."

"Sorry," the cold voice on the other end of the line said. "We gotcha. Be at Gate C 19 in thirty minutes."

"Son of a bitch," I yelled as I threw my phone on the seat beside me. I felt tears of anger and disappointment welling up in my eyes. Lee and I had plans this evening for dinner at Canopies, my favorite restaurant in Panama City, and we were supposed to meet several friends there as well. I had not seen him in almost a week, and now he would have to go to dinner alone, and I would be lucky to get something from a vending machine. This was only one example of how crew scheduling operated.

Cell phones were not as sophisticated in 2001 as they are now, and sometimes it was tricky to see who was calling you. After that incident, I learned from the pilots, and as soon as I got off the airplane at the end of a trip, I ran and got out of uniform, stopped by a bar in the airport and ordered a beer or cocktail. That way if crew scheduling called, I just told them "Yes, I'm in the airport, but I'm in the bar and I'm drinking. Sorry, but you will have to find someone else."

Even this didn't always work, because sometimes they called the cockpit as the plane pulled up to the gate and told the pilots to have me call crew scheduling. This really irritated me, because if you were on the aircraft you definitely couldn't say you were drinking, and if you didn't call them back you got an occurrence.

Crew scheduling also had the authority to drop and swap trips for flight attendants when I first started flying; later on, after ASA was bought out by Delta, our bids and trips were all done by computer, and so if you wanted to swap a trip with someone, or post a trip you did not want for someone to pick up who needed extra money, life became a lot easier.

In the first years with ASA, our bids and trips were done on paper, and any changes, drops, or swaps of trips had to be approved through crew scheduling. They were notoriously slow with this process, and sometimes by the time you found out if you had been able to drop a trip, it was no longer even worth it because you couldn't make any plans until it was too late and you were already on the way to the airport to duty in for work. No one had anything good to say about crew scheduling....ever.

Ask any flight attendant or pilot who ever worked for ASA/Delta Connection . Crew scheduling is the devil.

## 24 Hang-Glider

Charles "Sully" Sullenberger was a glider pilot; I will believe forever that God put him in the seat of his aircraft on the day he was destined to land in the Hudson River, just as sure as I am sure that the hang-glider who almost hit the windshield of our regional jet as we were leaving New York will forever be a believer in Divine providence and intervention.

Sully was quoted as saying to Katie Couric, "One way of looking at this might be that for 42 years, I've been making small, regular deposits in this bank of experience, education and training. And on that day the balance was sufficient so that I could make a very large withdrawal."

I had resigned my position with ASA/Delta Connection approximately four years before Sully's heroic ditching of US Airways flight 1549 in the Hudson River in which all of the 155 passengers and crew aboard were saved, and as I watched the news coverage of the event from my living room, I cried and cheered.

One of the more interesting elements of Capt Charles "sully" Sullenberger's flying experience is his training flying gliders. Many believe that his experience and familiarity flying such engine-less aircraft could have played a large role in his saving of Flight 1549 when he lost both engines to bird strikes on that wintry day in New York.

Although most pilots never have an experience quite like Sully's in their careers, many have performed quietly heroic maneuvers which never make the news, but which are life-saving and require great skill , a calm, cool head, and the ability to make quick decisions under pressure.

One has to believe that after hundreds of takeoffs and landings, that eventually, not matter how safe air travel is, the law of averages might catch up , but no one really talks about it much. After 9/11, I had a small white teddy bear, given to me by a New York firefighter, which I attached to my flight bag handle. The bear became my good luck charm, and once when I accidentally left it at home, I was nervous the entire trip without it.

We had just taken off from White Plains, New York on a beautiful fall morning, and I heard the four tones indicating we had reached 10,000 feet. I unfastened my seatbelt and stood up in the galley to start the coffee, glancing through the small window as I stood. Suddenly the plane began to lose altitude at an alarming rate, and

I sat back down in utter fear. I knew something was wrong. My heart was beating so loudly that I could hear it in my ears, and I began to pray while at the same time trying not to show any alarm in my face to the passengers. Several of them had made audible gasps and loud comments, but as suddenly as the plane had seemed to make a dive, it began climbing again, and then leveled off.

"Ding-dong," I heard the chimes which signaled, "cockpit calling."

I quickly lifted the phone , and listened for the Captain's voice.

"Sorry about that," he said. "We are fine, but just had a very close encounter with a glider pilot."

"Oh, my God, " I exclaimed. "Is everything ok?"

"Yep, but we saw the whites of his eyes, and I'm sure he's going to have to change his pants when he gets home," laughed the Captain. "So are we," he added.

"I'll make an announcement in a bit to the passengers," he said, "but I want to think about exactly what I want to say. No need to panic everyone by telling more than we need to."

"I think that is the worst feeling I have ever had since I have been flying," I admitted to the Captain.

"Everything is fine, don't worry," he assured me. "I'm sure it was scary feeling the loss of altitude right after takeoff and not knowing what was happening."

I hung up the phone and got back up to start preparing the service cart in the galley. I set two Crown Royals on the top shelf and took eight dollars out of my purse to pay for them. I had plans for that whiskey as soon as I arrived at my hotel room in Myrtle Beach, our final destination of the day.

I began pushing the cart to the aft of the aircraft to begin the beverage service. My hands were shaking.

## 25 Commuting

In general, I do not like the Atlanta gate agents, especially those who work for ASA/Delta Connection; I am sure that many of them are probably nice people, but I had very few pleasant experiences while trying to commute out of Atlanta to Panama City.

"Have you called the standby's yet?" I asked softly, making sure that the gate agent was not busy with any of the ticketed passengers." I literally felt as though I were walking on eggshells anytime I had to ask these gate agents a question.

"I'll call you if there is a seat available," she answered curtly, making no eye contact.

*"Great,"* I thought. *"Another Miss Congeniality working the Panama City flight,"* or PFN as the airport code was listed at that time.

At present, Delta Connection has flat-screen monitors at each gate which lists standby and cleared passengers, but when I was flying (2000-2005), the screens did not exist, and gate agents held the power. Their friends, their family, or someone with a buddy pass could bump you off a flight, even if your seniority number was higher…..especially if you were a flight attendant. Many of the gate agents had applied to be a flight attendant and not gotten the position; therefore, if you did anything at the gate to slightly piss them off or annoy them, you could sit in Atlanta for hours, until all of the flights taking you home were departed. Pilots were a little luckier than flight attendants because they always had the option of sitting in the jump seat, the fold-down seat in the cockpit used by FAA and sometimes "deadheading" pilots, those who were being flown to another location to pick up a flight.

According to policy, the gate agents can let you know forty-five minutes to 10 minutes before the aircraft departs if there is a seat available, but I could already tell by the smug, hateful expression on this gate agent's face that if I got a seat at all, it would be at the last minute. Sometimes, even if there was a seat, they would "forget" to call you if they became angry; therefore, I became quite adept at keeping my mouth shut and trying to be pleasant to gate agents, no matter how unpleasant they were in return.

Several other stand-by passengers were waiting for seat assignments at the gate, and one of them was a Delta pilot who had been pretty persistent in his request for information.

"I told you the jump seat is already taken on this flight by one of the ASA pilots," she snapped. "I will let you know if there is a seat as soon as I can. "

"May I help you ma'am," she was asking the woman in line behind the Delta pilot, who was a ticketed passenger. The pilot moved slightly to the side and continued to stand at the counter, glaring at the gate agent.

We were now twenty minutes before the flight was to take off, and the gate agent had started the boarding process, completely ignoring the stand-by passengers. Soon all of the ticketed passengers had boarded, and I had carefully counted each one as they passed through the door leading to the gate. Only forty-one passengers were aboard, and I knew there were sixty-six seats on the ATR. The gate agent continued to stare down the Concourse as though she were expecting a stampede of late-arriving passengers. I couldn't stand it any longer. I approached the counter again.

"Excuse me," I began, "but we are almost at the ten minute cut-off for boarding, and you still haven't cleared the stand-bys. There are less than fifty people aboard, and I know there are sixty-six seats."

"I will let you know when stand-bys are cleared, if you will just be seated," she snapped.

"Fine, " I snapped back. " but I prefer to stand, and I will be seated when I am on the plane. " I stood glaring right back at her. The other stand-bys apparently gained some confidence as they listened to my refusal to back down and they stood up as well.

Seven of us stood glaring at the gate agent, and finally she announced that stand-bys were cleared and printed out our seat assignments.

I snatched mine from her hand as I passed through the door, and as soon as I was out of earshot I bravely shouted, "Bitch!" I knew no one could hear me over the loud noises on the busy tarmac area surrounding the aircraft. As much as I disliked the ATR, it always looked beautiful to me when it was taking me home after a long trip.

When you work for the airline, you are permitted to live in other cities than the one you are "based in," but it is completely your responsibility to make it to the base on time. If you can't get a standby seat from your hometown, you must leave in time to drive. Living in Panama City came along with the most difficult commute imaginable because we are a Spring Break, summer vacation, winter vacation, Iron Man, Triathalon city, along with many more events and festivals which keep the flights very full. Additionally, the runway at that time was too short for the larger jets, and so the only flights going in and out of Panama City were 50 and 66 seat aircraft, primarily ASA and Northwest. Northwest Connection, at that time, flew only to Memphis from Panama City, and so even if they had an empty seat, you had to gamble on whether or not there would be a seat from Memphis to Atlanta, and by that time it was just as easy to make the four and a half hour drive to the Atlanta Airport.

Most monthly schedules for flight attendants usually amount to between sixty-five to one hundred hours of flying, and at the time I started we were required to fly at least sixty-five hours, usually divided into three and four day trips. When all is said and done, most flight attendants work about two weeks a month; however, if you think about how tired most people are after one full day of flying, it is easy to imagine how much stress that two weeks per month of flying puts on the body as well as the mind.

Many flight attendants and pilots who work for Delta live in the

Atlanta area, or in the outlying areas such as Newnan, Marietta, or Peachtree City, but just as many commute from other cities. Getting to the airport on time is just as risky living in or near Atlanta as it is flying from another city because of traffic congestion. I usually flew to Atlanta the night before I started a trip, slept at the crash pad in Stockbridge, and then still stressed out about the drive to the airport which normally should take about twenty minutes, but if there was any problem with traffic could take two hours or more. There was always plenty of time to incur a trip failure for missing a trip; sometimes it was just in the cards to fail as a flight attendant no matter how diligent and careful a person was. Delta is much more flexible and forgiving of this type of thing than ASA was; many of those flight attendants who survived at ASA through the transition to Delta Connection and better times are still flying today, including my former crash pad roommate Valerie, as well as Pam, my friend who lives in Panama City who first encouraged me to come to Atlanta and become a flight attendant.

Today there is a new airport in Panama City, Northwest Florida International (ECP), and commuting is not as difficult as it once was because the larger MD90 aircraft can land there, and Delta Connection no longer uses the ATR and Brasilia aircrafts.

Margo Deal Anderson

## 26  Stowaways

As the ATR began to roll, I glanced back in the cargo area where two of the pilots from Tri-Cities were lying on top of their bags and those of the passengers, grinning at me as I finished my announcement and began the walk up the aisle to do final checks.

*"God, why do I do this to myself,"* I thought, referring to the conspiracy I had entered into with the other three crew members in allowing three pilots to commute home in the baggage compartment because there were no available seats.

Spring Break was a ridiculous time of year to be a commuter, and I had been left behind in Atlanta many times watching the aircraft fly away that could have had me home in less than an hour. The pilots would not be any trouble; they would sleep all the way there, and I didn't have to worry about them telling anyone because they would have been fired just as surely as I would have.

Things were more relaxed in the blissful days before 9/11, and in truth, I always liked the idea of having extra pilots around.

Valerie and I had bid a trip together this month, and we had actually been lucky enough to get it. How wonderful to fly with someone familiar, someone who would do her share of the work, and someone I could joke around and have fun with. The two of us traded back and forth the jobs of Flight Attendant A and Flight Attendant B, and Valerie actually preferred the "B" position because she didn't like making the announcement. This was fine with me, because I didn't like sitting in the front with everyone staring at me. Perfect.

We were flying to Tri-Cities today, and this was one of the most difficult commutes for some reason. The Tri-Cities airport services northeast Tennessee and southwest Virginia, and the flights are always full, thus the stowaway situation. As I finished the announcements, Valerie brought the paperwork back, and just as I began a walk -through to make sure all the carry-on baggage was stowed properly, one of the pilots in the back said something.

I pulled the baggage curtain back, and peered into the partial darkness at the three of them.

"Did somebody need something?"

"Sorry to bother you, Margo, but do you have time to pass back a can of Diet Coke?" Carl was a really nice guy, and he had such a baby face that most passengers who saw him wanted to verify that he was actually old enough to fly the plane.

"Sure, no problem," I said. "Do you want a cup of ice? The canned drinks are not cold."

"That would be great," he replied.

I quickly filled a cup with ice, grabbed a Diet Coke out of the drawer and passed it back to Carl before pulling the curtain shut.

Soon we were at altitude, and Valerie and I were pulling the beverage carts out to start serving, when Carl, followed by another pilot, Steven, began climbing out of the baggage compartment. Carl had to go to the bathroom, and Steven wanted to go to the cockpit to see the two pilots who were actually *flying* the plane.

Steven went up the aisle ahead of us so that he would not get in the way of our beverage service, and by the time we started serving the first row in the front of the aircraft, he had already disappeared into the cockpit.

"Was that one of the pilots?" asked the lady in 2C. "Isn't he supposed to be helping to fly the plane?"

"He had to go to the bathroom," I said.

"I never saw him come out," she replied. "When did he come out?"

"Oh, I don't know," I said airily. "You were probably looking away and he was probably in a hurry."

Valerie snapped the brake on the cart and we quickly moved to the next row of seats before the lady could ask another question.

As we were serving at Row 7, I glanced back at the galley area and saw Carl walking around munching on Biscotti cookies, and Peter coming out of the head talking with a passenger who had gotten up to use the bathroom, probably trying to explain how there could be two pilots in the galley instead of in the cockpit.

I quickly looked away and tried to keep my mind on the beverage service instead of on the three pilots who couldn't seem to keep their asses in the baggage compartment out of sight.

When we finally completed the beverage service, Steven came out of the cockpit, and as he made his way to the aft of the aircraft where Valerie and I were staring at him in amazement, he stopped along the way making small talk with various passengers.

"Great," said Valerie. "Why can't they stay out of sight? Is it just an ego thing with pilots that they imagine they can get away with anything, or are they just stupid?"

"Hey," said Carl. "Watch it now!"

"Sorry," said Valerie. "Just kidding, but honestly, can you guys be just a little more discreet about this?"

Soon they were all safely in the baggage compartment again, and our new worry was how to get them out and off the aircraft before the baggage guys opened the galley door and started unloading the baggage but at the same time waiting for the other passengers to exit first without seeing all of them.

"I am not doing this anymore," I said to Valerie. "My nerves will not take it."

"I know; I'm going to have to start taking a Xanax before every flight just to deal with this kind of crap," she laughed.

We finished cleaning up the galley, and Valerie walked through the aisle with the "thank you" bag collecting all of the trash. When she returned, an arm stuck through the cargo curtain with two empty drink cans , a handful of napkins, and empty pretzel bags.

I snatched the items from the hand and slapped it back inside the compartment, giggling the entire time.

The four tones sounded indicating initial approach, and mercifully, the flight was almost over. Soon we were touching down in Tri-Cities, and as soon as the aircraft began to park at the gate, I dragged the stowaways out of the compartment, stuffed two of them in the head, and pushed one against the galley door behind me out of sight of the exiting passengers.

About half of the passengers had deplaned and there was a lull in the line allowing me time to shove Carl into the line just before the galley door opened behind me where he had been standing.

"Hi," I shouted to the baggage guy who climbed up into the galley and began throwing bags down to his partner below.

After the last passenger had deplaned, I opened the door to the lav and said, "Coast is clear; come on out!"

"Thanks so much, Margo," said Steven.

"Yeah, thanks for rescuing me out of the head," laughed Peter. "If you had left us in that tight space much longer I was afraid Steven was going to kiss me."

Steven punched him so hard they both almost fell down the stairs, but at least they were gone.

I sat down on the jump seat and breathed a deep sigh of relief. I was still employed.

Margo Deal Anderson

## 27  TSA, Underwire Bras, and Red Tennis Shoes

So many things changed about flying for both the flight crews and the passengers after 9/11, but TSA quickly replaced crew scheduling as the most despised organization in the airline world.

The days of simply showing an airline ID identifying one as a flight crew member and having access to the tarmac and aircrafts had come to an end.

My first experience with the TSA organization was in Asheville, North Carolina, when as I walked through a metal detector the alarm went off.  The agents were brand new at their jobs, and I think they were as nervous about the whole thing as we were. The young man spoke sharply to me and said, "You will need to step over here," as he pointed to an area over to the side.  I obligingly stepped over to the side where he began to move his wand over my body waiting for it to alert him to whatever I had on me that was metal.

"Maybe it's the metal on my ID's," I offered helpfully.

"Would you step through the metal detector again," he snapped curtly.

"Well, ok. You don't have to be so rude," I retorted, as I watched my bag, my lunch, and my purse disappear on the rolling counter moving through an x-ray machine.

"You will have to remove your shoes and your sweater too," the agent sitting in the chair watching the x-ray machine said.

"My shoes? You have to be kidding me."

"Remove your shoes, your sweater, and place everything on the counter to go through the x-ray machine," the agent repeated with no expression.

I took off my shoes and my sweater and threw them on the rolling belt, and the Captain and First Officer did the same. They also had to remove their wallets, belts, and hats to go through the machine. They were not happy either.

When I walked through the metal detector for the third time I continued to set off the alarm.

"Now what do I do?" I asked as I glared at the agent.

"It must be an underwire in your bra," he said.

"You've got to be kidding me; what am I supposed to do now, take off my bra?"

"I'm afraid so."

"Really." I said it as a statement not a question.

The agent continued to stare at me as I glared back at him.

"Fine!" I shouted. I knew how to take off a bra without removing my shirt. I did it every day of my life on the way home from work. I reached behind me and undid the hooks, reached up my right sleeve and puled the strap down, did the same for my left sleeve, and then pulled the entire pink satin Bali bra out from under my shirt and threw it unceremoniously on the rolling belt and walked through the metal detector. It did not go off this time.

On the other side of the machine, I retrieved my shoes, put them on, and then ran one of my stockings as it caught on the chair I had sat in.

"Great," I thought. Now I will have a run for everyone to stare at all day long. I snatched the bra off the counter and carried it in my hand as we proceeded through the concourse to our gate and to the aircraft.

"Are you going to carry that around all day?" quipped the First Officer.

"Bite me," I said, and walked past him to the stairs leading to the aircraft.

I did my safety checks, made the coffee, and looked at the passenger manifest, while the crew did their own work. The passengers were headed to the aircraft by now, and I hastily stuffed my bra into the snack drawer before greeting the first gentleman as he boarded the aircraft.

"Good afternoon sir; welcome aboard," I smiled pleasantly, wondering if he had been frisked and separated from his underwear too.

"Hello," he smiled as he moved down the aisle to his seat in the sixth row.

I was still absolutely furious at the nerve of the TSA agent, and I wondered if we were going to have to go through this procedure every time we boarded a flight, or if this was just temporary because of 9/11.

The passengers were unusually grumpy as they entered the aircraft, and several made comments about the rudeness of the TSA agents, expressed disdain at having to remove their shoes, and complained about the flight boarding over thirty minutes later than scheduled.

I couldn't tell them why we had delayed the boarding, but after 9/11, flight crews pulled out every seat cushion, double-checked all of the overhead bins, checked all of the seat pockets, and other possible places that weapons could be concealed on the airplanes. Most flight crews believed that the box cutters used to murder the flight attendants and others were placed on the doomed American and United flights by food service or cleaning personnel who had access to the airplanes. We took a lot of abuse from gate agents and flight supervisors during that time about delayed boarding of passengers, but we were the ones who were going to be alone in the skies once the doors closed to the airplanes, and we did everything possible to assure the safety of ourselves and our passengers. The screening process for everyone who has access to aircraft is not the same, and aircraft still remain vulnerable even now, years after 9/11.

I also became more and more dismayed at the lack of security in two crucial areas. In the first weeks following 9/11, security was very tight, even for employees, but this was soon relaxed. Employees enter the employee parking lot at Hartsfield Airport, and after showing ID's at the gate, the luggage and other items in employee cars are not checked; after parking their cars in the employee lot, these employees board shuttle buses to the employee work areas, flight attendant and pilot lounges, and duty-in stations. In other words, an employee could bring in a tool chest or carry-on luggage, and then after working all day, decide to go upstairs and board a flight carrying the bag or item with them to any destination, never having the item subject to any type of security before boarding. Items could also be placed in the baggage compartment of any aircraft brought in by these employees with this type of access. Even today, when the public is subjected to every possible scrutiny by TSA, this is a cause for concern, and in my opinion, even alarm. While it is true that airline employees are subject to intense screenings and security checks when hired, and the large majority of employees are completely trustworthy, in today's climate of radicalization and lone wolf attacks, the public needs to be aware of this weakness in airport security.

The second area of major concern for me at the time has been corrected now; while passenger carry-on bags were put through security checks, baggage which was checked was not screened. This lack of screening went on for many months after 9/11, but now seems to have been corrected, with checked bags and packages also undergoing x-ray and other types of scrutiny by airport security.

The flight from Asheville to Atlanta was uneventful after boarding and takeoff; Asheville passengers are usually what I call "easy." Most of them drink fruit juices or tea, they read or sleep in flight, and they are very conscious about trash and keeping everything clean. I was always amazed at how each city seemed to have its own passenger "personality" traits. Myrtle Beach golfers were rude and demanding, Panama City and Ft. Walton Beach were always party flights, and White Plains, New York was a messy, loud, demanding group.

We had a two-hour break in Atlanta, giving us time for lunch before we departed for Augusta and a long layover at the Partridge Inn. Tomorrow would be an early day, and once we left Augusta at 11:00 a.m. and returned to Atlanta, I would be finished with the trip and ready to head home to Panama City.

Halfway through lunch, I remembered that I had left my underwire bra in the snack drawer of the aircraft, but it was long gone with a new crew by now.

"*Great,*" I thought. "*I'm sure that crew will wonder what in the hell we were doing on the last flight.*" I laughed out loud thinking about it.

The pilots were still kidding me about the underwire bra the next morning as we were leaving Augusta, when the security agents asked both of them to remove their shoes. Shoe removal was a very new practice, and while it feels normal in today's procedures, the process felt extremely invasive at first. I thought the pilots were going to come unglued. There was a large group of passengers leaving the airport at the same time we were boarding, and somehow when the pilots began retrieving their belongings, the Captain's shoes were gone.

"Where in the hell are my shoes?" he demanded of the security agent.

After ten minutes of searching , the shoes were nowhere to be found.

"How is this even possible," the Captain shouted. "This is one of the smallest airports in Georgia; the shoes couldn't have just disappeared."

"Maybe someone took them by mistake," I giggled, thoroughly enjoying the situation.

The Captain ignored me and declared that it was against regulations for him to fly without shoes.

"Don't you have any extra shoes with you?" asked the First Officer.

"No, " he shouted.

The flight was delayed while we took the airport shuttle to the Augusta, Georgia K-Mart store. The Captain walked inside in his socks, and he emerged fifteen minutes later wearing red high-top tennis shoes .

I was hysterical all the way to the airport, and I never stopped smiling for the entire flight.

Margo Deal Anderson

## 28 Astronauts, Ballerinas, Teachers, and Nurses

For as long as I could remember I dreamed of being an astronaut. My dad was with the Corps of Army Engineers and was part of the space program from 1963 until 1969 when we lived at Cape Canaveral, Florida. I was in the third grade when I saw my first launch from my bedroom window on Merritt Island, Florida, as the roaring of the rocket engine shook the house and I stood on tiptoe on my bed peering out of the windows of our brand new house in Bel-Aire Subdivision.

The house was brand-new, with terrazzo floors, tile bathrooms, central air-conditioning, a patio for barbeques with neighbors, four orange trees and two grapefruit trees growing in the backyard, and an ultra-modern galley kitchen, complete with dishwasher and garbage disposal. My parents had paid a whopping $17,000 for this three-bedroom home in a beautiful little subdivision where most of the fathers worked with NASA or some part of the space program, and every kid had a stay-at-home mom. The early 1960's were a magical time to be in elementary school; science was my favorite subject, and I could not wait to fly into space. I was sure that someday I would explore the moon. That dream did not last long.

"I am going to be an astronaut when I grow up," I declared solemnly to the rest of my third grade class at Tropical Elementary School. There was a moment of silence before several of the boys laughed.

"You can't be an astronaut," laughed David Earhart, my biggest crush at the time, which made the laughter hurt all the more.

"Why not?" I challenged as I stood trembling in front of the class during our weekly "show and tell" session. I had brought a model of a spacecraft, along with my Barbie and Ken dolls.

"Because girls can't be astronauts," he said knowingly. David Earhart was a distant relative of Amelia Earhart, and he never let anyone forget it.

"I'm afraid he's right," agreed Miss Motes. There was a sadness to her voice as she continued. "Little girls can grow up to be teachers, nurses, or ballerinas, but not astronauts. You have to be a military pilot to qualify for the astronaut program, and girls are not allowed to be pilots in the military."

I began to cry, and I ran out of the classroom, shouting, "It's not true, that's a lie..." I ran all the way to the row of bicycles, and I pulled mine from the bicycle rack and began riding it home on the paved bicycle path which followed the road from my subdivision all the way to the elementary school, about a mile away. When I arrived home, I got into trouble for leaving school, and my mom sent me to my room to wait until my dad got home from work, leaving punishment to him.

This was the beginning of a long list of disappointments I would face as a little girl growing up in the 1960's, but my love of science and the dream of flying did not go away. My best friend, Cheryl Chappell, who was two years older than I was, and whose father worked with mine, flew twice each year to spend time with her grandparents in Oklahoma. I rode with her to the airport one time and waved goodbye to her with her parents as she boarded the sleek, beautiful Eastern airlines jet with the blue and turquoise stripes on the side. I could only imagine how wonderful it must have been to be inside as it left the ground, disappearing out of sight just a few moments later.

"I want to fly in an airplane like Cheri," I declared after watching my friend fly all by herself to visit her grandparents. "Flying is the most wonderful thing in the world."

I would like awake at night dreaming of flying an airplane, imagining what the inside of the cabin would look like, and hoping for the time when I would have the chance to actually go somewhere on an airplane.

While we were living at Merritt Island, Florida, my family often drove to Panama City, Florida to visit our relatives, and in the 1960's , before the opening of Walt Disney World, Orlando and Central Florida was a swampy barren area forty miles east of us where we would drive to get on Interstate 10 and head north. Before one of these trips my Dad had a surprise for me. They had arranged to drive me to the airport in Orlando, where I would fly on National Airlines, all by myself, to Panama City. Relatives would meet me there and my parents and younger brother would see me that evening after driving the eight-hour distance. I was excited beyond words; I was finally going to fly!

I was mesmerized by the two beautiful flight attendants on the National Airlines flight; they took me to the cockpit to meet the pilots and then spoiled me for the entire hour long flight with soda and snacks.

I stared at their beautiful makeup, their elegant French twist hairstyles, their short skirts, seamed stockings, and glamorous high heels, and, of course, their dazzling smiles. I forgot all about being an astronaut and decided that I would be a flight attendant instead.

To my father's dismay, I made the announcement of my new career choice as soon as we were reunited in Panama City, and over the years I became more and more determined that this would be my chosen career path. During the late 1960's and 1970's the uniforms became even smaller and more provocative, and the lifestyle of the flight attendants more mysterious and Hollywood-like. When I was a senior in high school, my father convinced me that pre-law was a much better choice, and off to the University of Southern Mississippi I went as a pre-law major, eventually finishing a degree in language and literature.

Twenty-seven years and two husbands later, at age forty-five, I rebelled against my tedious job as a teacher, against a boring small-town life, and flew to Atlanta, Georgia to interview for a position as a flight attendant. Despite many of the negative aspects of the job, and the fact that 9/11 happened less than a year after I started, I made some of the best friends of my life, had the most fun I've ever had, and my life is richer for the five years I spent as a flight attendant for ASA/Delta connection.

# 29 Epilogue

My flying days with ASA ended abruptly in the fall of 2005, when I called my inflight supervisor in Dallas from Panama City to tell her that I was sick with sinus and ear infection and would not be able to work my upcoming four-day trip. At this time I was flying part-time, and I had been calling in quite often; in fact, I was hardly flying at all. I was sick of it, my ears caused me extreme pain if I had the slightest cold, and I used almost any excuse not to go.

As I spoke to her there was deafening silence on the other end of the telephone line; she made no response as I told her that I could not fly because of the ear infection and how painful it would be for me.

She let me finish my story and then informed me curtly that I had been absent too often, and that if I did not arrive in Dallas to take my scheduled trip that I needn't ever come back. I reminded her that the "piece of crap" aircraft of ASA which lost pressure was the reason my ears were damaged in the first place, and that I could document with my doctor that I had not had any problems with my ears until the decompression incident aboard their aircraft.

She repeated to me again that if I didn't show up for my four-day trip that I needn't ever show up again; I took a deep breath and said, "Fine then; I am finished. You and Atlantic Southeast Airlines can go to Hell." I hung up the phone and burst into tears. My flying career had ended just as suddenly as it had begun five years ago. I sat on the bed in the loft of our lake house and held Lamb Chop as I looked out the large glass windows toward the blue waters of the lake, petting him as I thought about what I had just done.

"Well, Lamb Chop; I am home for good; no more little black flight bag for me, and no more staying in the kennel for you," I declared.

I went downstairs to the kitchen with Lamb Chop following close behind, and I poured myself a Crown and Diet Coke.

"*It's nine a.m. and I am drinking. Great.*" I said to myself. I went out on the back deck overlooking the lake, sat down in one of the comfortable lounge chairs and finished the drink before I called Lee to tell him that I was no longer employed by ASA/Delta Connection.

"Why don't you sue them," he asked. "It's not your fault your ears are damaged; their piece of shit airplane caused you to have the problem, and then you got worse when they wouldn't allow you to take sick leave. Sue the bastards!"

"I don't think so," I said. "I would probably just lose, and it would be a long, expensive, stressful process trying to fight an entire airline."

" I will support you in whatever you want to do," Lee said gent[ly]. "It will be kind of nice having you home all the time," he added.

"Maybe I will write a book," I said. " Maybe I will write a book [that] tells everybody what it is really like to be a flight attendant [for an] airline like ASA.."

Memoir of a Flight Attendant

## ABOUT THE AUTHOR

Margo Anderson became a flight attendant in March of 2000 on a whim, a suggestion from her dear friend, Pam who is a flight attendant with Delta Connection to this day. Margo has a Master of Arts in Language and Literature, and she has always kept journals; the journals she kept as a flight attendant are the basis for this memoir.

During her five years of flying she wrote in a journal almost daily, recording snapshots of the lives of flight attendants, pilots, baggage handlers, gate agents, maintenance crews, and of course, the passengers.

She also works as a staff training consultant in the educational field, sings professionally, and is currently the Mayor of the town where she lives, Lynn Haven, Florida. Margo has one daughter and two granddaughters, and she is married to the love of her life, Lee Anderson, a combat veteran of the Vietnam Conflict, who has four children and eleven grandchildren.

They travel often, almost exclusively with Delta Airlines; Margo admires flight attendants to this day, believes that the flight attendants who work with the commuter airlines are the hardest workers in the entire airline industry, and she is most happy to see that Delta has returned to some of the earlier glamour of the flight attendants of her childhood with their newest uniforms.

They enjoy spending time in the beautiful city where they were married, Key West, Florida, and Margo is currently working on her second book, *The Mayor*.

Made in the USA
Charleston, SC
13 April 2016